Bioinformatics
Database Systems

Bioinformatics Database Systems

Kevin Byron
New Jersey Institute of Technology, Newark, USA

Katherine G. Herbert
Montclair State University, New Jersey, USA

Jason T. L. Wang
New Jersey Institute of Technology, Newark, USA

CRC Press is an imprint of the
Taylor & Francis Group, an **informa** business

A CHAPMAN & HALL BOOK

CRC Press
Taylor & Francis Group
6000 Broken Sound Parkway NW, Suite 300
Boca Raton, FL 33487-2742

© 2017 by Taylor & Francis Group, LLC
CRC Press is an imprint of Taylor & Francis Group, an Informa business

No claim to original U.S. Government works

Printed on acid-free paper
Version Date: 20161122

International Standard Book Number-13: 978-1-4398-1247-1 (Hardback)

Library of Congress Cataloging-in-Publication Data

Names: Byron, Kevin, 1954- , author. | Herbert, Katherine G., author. | Wang,
    Jason T. L., author.
Title: Bioinformatics database systems / Kevin Byron, Katherine G. Herbert,
    Jason T. L. Wang.
Description: Boca Raton : Taylor & Francis, 2017. | Includes bibliographical
    references and index.
Identifiers: LCCN 2016027400 | ISBN 9781439812471 (hardback : alk. paper)
Subjects: LCSH: Bioinformatics. | Biology--Databases.
Classification: LCC QH324.2 .B97 2017 | DDC 570.285--dc23
LC record available at https://lccn.loc.gov/2016027400

Visit the Taylor & Francis Web site at
http://www.taylorandfrancis.com

and the CRC Press Web site at
http://www.crcpress.com

Contents

List of Figures

List of Tables

Preface

At some time during the course of any bioinformatics project, a researcher must go to a database that houses biological data. Whether it is a local database that records internal data from that lab's experiments or a public database accessed through the Internet, such as NCBI's Gen-Bank, researchers use biological databases for multiple reasons.

One of the original reasons for initiating the fields of bioinformatics and computational biology was the need for biological data analysis. In the past several decades, biological disciplines, including genetics, molecular biology and biochemistry, have generated massive amounts of data that are difficult to organize for efficient search, query and analysis. If we trace the histories of both database development and the development of biochemical databases, we see that the biochemical community was quick to embrace databases. For example, E. F. Codd's seminal paper, "A Relational Model of Data for Large Shared Data Banks" published in 1970 is heralded as the beginning of the relational database, while the first version of the Protein Data Bank (PDB) was established at Brookhaven National Laboratories in 1972.

Since then, and especially after the launching of the human genome sequencing project in 1990 [217], biological databases have proliferated, most embracing the World Wide Web technologies that became available in the 1990s. Now there are thousands of biological databases, with significant research efforts in both the biological and the database communities for managing these data. There are conferences and publications solely dedicated to the topic. For example, Oxford University Press dedicates the first issue of its journal *Nucleic Acids Research* (NAR) every year specifically to biological databases. In 2013, this 20th annual database issue exceeded 1300 pages and included 88 new online molecular biology databases. Among the new databases presented is one containing transcriptome profiling (RNA-seq) data for monkeys, providing new cellular mechanism information about our closest relatives. The NAR database issue is supplemented by an online collection that lists 1,512 databases in 14 categories and 41 subcategories, including both new and updated ones. In addition, Oxford University Press publishes the open access journal, *DATABASE: The Journal of Biological Databases and*

Curation, which contains articles describing novel biological databases and software tools designed to interact with these databases.

Biological database research now encompasses many topics, such as biological data management, curation, quality, integration and mining. Biological databases can be classified in many ways, from the topics they cover, to how heavily annotated they are or which annotation methods they employ, to how highly integrated their databases are with other databases. Popularly, the first two categories of classification are used most frequently. For example, there are archival nucleic acid data repositories (e.g., GenBank) as well as protein sequence motif and domain databases that are derived from primary data sources.

Modern biological databases comprise not only data, but also sophisticated query facilities and bioinformatics data analysis tools; hence the term "bioinformatics database systems" is often used. In this book, we endeavor to explain the world of bioinformatics database systems.

The book is divided into seven chapters. Chapter 1 summarizes the popular and innovative bioinformatics repositories currently available. Discussions include popular primary genetic and protein sequence databases, phylogenetic databases, structure and pathway databases, microarray databases and databases that do not fit into the aforementioned categories, also known as boutique databases.

Chapter 2 discusses the data quality and information integration issues currently involved with managing the bioinformatics databases. With many of the bioinformatics databases looking toward interoperability between the databases, issues concerning data quality are arising. It has been said that the relationship between data quality and data standards is not clearly understood. Do data standards necessarily increase data quality? Depending on the type of data standard and the aspects of data quality considered, the answer to that question is variable. Providing for adequate data quality planning is crucial. Discussions include some of the data quality issues observed in the bioinformatics databases. This chapter presents efforts in the data cleaning field to help with the data quality problems in these databases.

Chapter 3 discusses biological data integration issues and demonstrates how data integration can create new repositories to address needs of the biological communities. The chapter also presents typical data integration architectures employed in current bioinformatics databases. The emergence of high-throughput genomic and transcriptomic datasets from different sources and platforms has enhanced our understanding of how these factors influence complex diseases. It is challenging, however, to explore the relationship between different types of such data sets. Tools available now allow a researcher to build his or her own integrated MySQL database from many popular data sources.

In Chapter 4, biological data searching is explored. Biological data exist in many forms. Macromolecules RNA and DNA are comprised of a handful to thousands of nucleotides. Proteins are formed from amino acids. When new RNA, DNA, or protein is encountered, searching through known elements is the first step in trying to learn about the new element. A search is performed based on the primary sequence, secondary structure or tertiary structure. To learn about evolutionary history of an unknown element, a phylogenetic tree search can be conducted in a database of known phylogenetic trees. If a biological sample is suspected of being diseased, searching through known diseased transcripts helps reveal important treatment intelligence.

Chapter 5 presents the topic of biological data mining. General data mining approaches are discussed, as well as specific data mining methodologies that have been successfully deployed in biological data mining applications. Open biomedical ontologies (OBOs), especially gene ontologies (GOs), are discussed in the context of establishing common vocabularies among large and growing numbers of information stakeholders. Two biological data mining case studies are highlighted that illustrate how data, query and analysis methods are integrated into user-friendly systems. The first case study presents a data mining methodology using a stochastic covariance model to predict the presence of evolutionarily conserved RNA secondary structure elements in a genome. The second case study discusses the genome-wide discovery of coaxial helical stackings. RNA tertiary motifs, such as coaxial helical stackings, are recurring substructures within non-coding RNA (ncRNA), and they play critical roles in a variety of nuclear and non-nuclear cellular functions.

Chapter 6 explores biological network inference. Next-generation sequencing (NGS) technologies generate precise information about DNA and RNA sequences. NGS provides much greater gene expression analytical potential as compared with older microarray technologies. As NGS technologies evolve and mature, a cell's transcriptome will be a reliable and highly detailed representation of gene expression. Using reverse engineering approaches, researchers hope to infer a gene regulatory network (GRN) that will help resolve many unanswered questions about cellular activity. Several current software tools and approaches for biological network inference are presented.

Chapter 7 presents biological data processing approaches using cloud-based technologies. The completion of the Human Genome Project at the turn of the 21st Century [217] and the subsequent advances in high-throughput sequencing of genomic and transcriptomic datasets [378] are increasing demands on the computational infrastructure needed for serious genomic research. As an example, modern high-throughput sequencers, such as the Illumina HiSeq series, produce sequencing reads from 150 bp in length up to a total of 120 Gbp of data per 27-hour run. Future medical advances depend on our ability to process and analyze large and numerous genomic and transcriptomic datasets. As an alternative to building and maintaining a local infrastructure, cloud computing, in many cases, offers on-demand access to flexible computational resources. There is an ever-growing recognition of the power of cloud computing for large-scale and data-intensive biological applications. We review the use made to date of cloud computing for the processing of biological data, and we discuss challenges presented for biological "big data analytics." The use of cloud computing technologies for bioinformatics research is still in its infancy. However, we anticipate rapid adoption of these technologies as more applications become available and cloud computing costs continue to fall.

Building bioinformatics database systems is a challenging but highly rewarding venture. This book provides enough background information and some state-of-the-art techniques in nascent cross-disciplinary fields. The book is targeted toward researchers and developers of bioinformatics database systems. It can also serve as a supplementary text for a one-semester upper-level undergraduate course or an introductory graduate bioinformatics course. We hope the book will inspire students, instructors, researchers and practitioners who use or build modern bioinformatics database systems.

Acknowledgments

We are grateful to Dennis Shasha for inspiring and bringing us to this exciting field, and to Shamkant Navathe for his support in publishing this book. We would like to thank our colleagues and students who have worked with us in all aspects related to this book, especially Miguel Cervantes-Cervantes, Zhang-Zhi Hu, Lei Hua, Christian Laing, Yongkyu Park, Nihir Patel, William H. Piel, Tamar Schlick, Huiyuan Shan, Bruce A. Shapiro, Yang Song, Junilda Spirollari, Dongrong Wen, John Westbrook, Cathy H. Wu, and Kaizhong Zhang. We also wish to thank Randi Cohen, Senior Acquisitions Editor of CRC Press, for her coordination and guidance, Sherry Thomas and Veronica Rodriguez, for their editorial assistance, and Robin Lloyd-Starkes, for her thoughtful comments on drafts of the book that improved its format and content. We are to blame for any remaining problems.

CHAPTER 1

Overview of Bioinformatics Databases

CONTENTS

1.1 INTRODUCTION

This chapter presents a variety of fascinating databases relevant to bioinformatics research. Shortly after the relational database was first introduced by E. F. Codd in 1970 [82], bioinformatics databases began to be established and made publicly available to researchers. Digitized biological information has been doubling every 12 to 18 months for years. Today's biomedical researchers are fortunate to have such an abundance of quality literature and information at their fingertips. Among the challenges for bioinformatics researchers (i.e., bioinformaticians), however, is trying to keep this deluge in some kind of order. Relevant databases are presented here and categorized as aids in understanding the resources that are available to researchers today.

TABLE 1.1: Sequence databases and their URLs

Database	URL
GenBank	http://www.ncbi.nlm.nih.gov/
EMBL-Bank	http://www.ebi.ac.uk/embl/
DDBJ	http://www.ddbj.nig.ac.jp
Ensembl	http://www.ensembl.org/
TAIR	http://www.arabidopsis.org/
SGD	http://www.yeastgenome.org/
GeneDB	http://www.genedb.org
dbEST	http://www.ncbi.nlm.nih.gov/dbEST/
Rfam	http://rfam.sanger.ac.uk/
RNA STRAND	http://www.rnasoft.ca/strand/
fRNAdb	http://www.ncrna.org/frnadb/
PIR	http://pir.georgetown.edu/
Swiss-Prot/TrEMBL	http://www.expasy.org/sprot/
UniProt	http://www.ebi.ac.uk/uniprot/

1.2   SEQUENCE DATABASES

This section centers on primary sequence databases. These are the most frequently used databases in the field. Attention will be paid to the large databases such as the National Center for Biotechnology Information's (NCBI's) GenBank [37] as well as its partners EMBL-Bank (Europe) [81] and the DNA Data Bank of Japan (DDBJ) [280]. The section also looks at species-oriented databases such as TAIR [258], and non-coding RNA sequence databases such as Rfam [61]. For the protein repositories, the section concentrates on UniProt [401] as well as its contributing data repositories, Swiss-Prot [279] and the Protein Information Resource [126, 174, 402, 403, 404]. Table 1.1 lists these databases and their respective URLs. In what follows we describe each of these databases in more detail.

1.2.1   Nucleic Acid-based Repositories

GenBank, EMBL-Bank and DDBJ

The most widely used biological databank resource on the World Wide Web is the genomic information stored in GenBank (USA) [37],

EMBL-Bank (Europe) [81], and the DNA Data Bank of Japan (DDBJ) [280]. GenBank is the National Institutes of Health (NIH) genetic sequence database, an annotated collection of all publicly available DNA sequences. Each of these three databases was developed separately, with GenBank and EMBL-Bank launching in 1980 [37, 81]. Their collaboration started soon after their development, and DDBJ joined the collaboration shortly after its creation in 1986. The three databases under the direction of the International Nucleotide Sequence Database Collaboration (INSDC) gather, maintain and share mainly nucleotide data, each catering to the needs of the region in which it is located [231].

Ensembl Genome Database

The Ensembl database is a repository of stable, automatically annotated sequences resulting from the Human Genome Project [217]. It is available as an interactive Website or downloadable as flat files. Ensembl annotates and predicts new genes, with annotation from the InterPro [269] protein family databases and additional annotations from databases of genetic disease (OMIM [22]), expression (SAGE [373, 392]) and gene family [113]. Since Ensembl endeavors to be both portable and freely available, software available from Ensembl is based on relational database models [177].

Arabidopsis Information Resource

The *Arabidopsis* Information Resource (TAIR) [258] is a comprehensive genome database that allows for information retrieval and data analysis pertaining to *Arabidopsis thaliana* (a small annual plant belonging to the mustard family). *Arabidopsis thaliana* has been of great interest to the biological community and is one of the few plants whose genome is completely sequenced [258]. Due to the complexity of many plant genomes, *Arabidopsis thaliana* serves as a model for plant genome investigations. The database has been designed to be simple, portable and efficient. One innovative aspect of the TAIR Website is MapViewer (http://www.arabidopsis.org/servlets/mapper). MapViewer is an integrated visualization tool for viewing genetic, physical and sequence maps for each *Arabidopsis* chromosome. Each component of the map contains a hyperlink to an output page from the database that displays all the information related to this component [258].

Saccharomyces Genome Database

The *Saccharomyces* Genome Database (SGD) [76] provides information for the complete *Saccharomyces cerevisiae* (baker's and brewer's yeast) genomic sequence and its genes, gene products and related literature. The database contains several types of data, including DNA sequence, gene-encoded proteins, and the structures and biological functions of known gene products. It also allows full-text searches of articles concerning *Saccharomyces cerevisiae*. The SGD database is not a primary sequence repository [76], but a collection of DNA and protein sequences from existing databases (GenBank [37], EMBL-Bank [81], DDBJ [280], PIR [404] and Swiss-Prot [279]). It organizes the sequences into datasets to make the data more useful and easily accessible.

GeneDB

GeneDB [166] is a genome database for prokaryotic and eukaryotic organisms. It contains genomic data generated from the Pathogen Sequencing Unit (PSU) at the Wellcome Trust Sanger Institute. The GeneDB database has four key functionalities. First, the database stores and frequently updates sequences and annotations. Second, GeneDB provides a user interface, which can be used for access, visualization, searching and downloading of the data. Third, the database architecture allows integration of different biological datasets with the sequences. Finally, GeneDB facilitates querying and comparisons of species by using structured vocabularies [166].

dbEST

dbEST [45] is a division of GenBank that contains sequence data and other information on short, "single-pass" cDNA sequences, or expressed sequence tags (ESTs), generated from randomly selected library clones. As of September 2012, dbEST contained approximately 73,360,923 entries from a broad spectrum of organisms, and the database keeps growing. Access to dbEST can be obtained through the Web, either from NCBI by anonymous FTP or through Entrez [326]. The dbEST nucleotide sequences can be searched using the BLAST sequence search program at the NCBI Website. In addition, TBLASTN, a program that takes a query amino acid sequence and compares it with six-frame translations of dbEST DNA sequences, can also be useful for finding novel coding sequences. EST sequences are available in the FASTA format from the /repository/dbEST directory at ftp.ncbi.nih.gov.

1.2.2 Non-coding RNA Repositories

Rfam Database

The Rfam database [61] categorizes non-coding RNAs and their conserved primary sequences and RNA secondary structures. A multiple sequence alignment (MSA) is used to represent each category. Each category also includes the consensus secondary structure annotation and covariance model (CM) for each MSA. CMs are produced using the Infernal suite of software [271]. Each non-coding RNA family consists of a set of RNA sequences believed to share a common ancestor. A representative alignment for the family (the *seed alignment*) is annotated with a consensus RNA secondary structure. A CM is built as a statistical representation of each family. The primary goal of the database is to provide a comprehensive and accurate set of non-coding RNA annotations for genome annotation.

RNA STRAND Database

The RNA STRAND (i.e., RNA secondary STRucture and statistical ANalysis Database) [18] is a curated and publically accessible collection of known secondary structures of many types and organisms. RNA STRAND provides the ability to browse, search, analyze and download any subset of its 4,666 entries. The Web-based interface allows a new entry to be submitted for review and addition to the collection. Each entry in RNA STRAND comes from another publicly accessible database and contains a link to its source database. Source databases include the Research Collaboratory for Structural Bioinformatics (RCSB) Protein Data Bank (PDB), tmRNA database, Rfam database, Nucleic Acid database (NDB) and others. RNA STRAND is useful for gathering RNA molecules that share specific attributes such as length range, source organism, NMR/x-ray validation, presence of a multi-loop, etc. Researchers studying unsolved RNA energy models, as an example, are able to analyze cumulative distributions and histograms for a certain group of RNA molecules using RNA STRAND. For further, deeper analysis and/or processing, the selected group of molecules can be downloaded in a variety of common formats.

Functional RNA Database

The functional RNA database (fRNAdb) [207], maintained at

http://www.ncRNA.org

is for comprehensive non-coding RNA sequences. fRNAdb helps in annotating non-coding transcripts acquired from publicly available databases. Each transcript is analyzed for various features such as maximum ORF length, the number of protein homologues, the average conservation score, transcription regulatory element motifs, existence of CpG islands, etc, that help in filtering out promising non-coding candidates.

1.2.3 Protein Repositories

Protein Information Resource

The Protein Information Resource (PIR) is an integrated public bioinformatics facility that supports genomic and proteomic research and scientific studies. PIR has provided many protein databases and analysis tools to the scientific community, including the PIR-International Protein Sequence Database (PSD) of functionally annotated protein sequences. The PIR-PSD, originally created as the Atlas of Protein Sequence and Structure edited by Margaret Dayhoff, contained protein sequences that were highly annotated with functional, structural, bibliographic and sequence data [126, 404]. PIR-PSD merged with UniProt Consortium databases [401]. PIR offers the PIRSF system [403] that classifies proteins based on full-length sequence similarities and their domain architectures, to reflect their evolutionary relationships. PIR also provides the iProClass database that integrates over 90 individual databases to create value-added views for protein data [402]. In addition, PIR supports a literature mining resource, iProLINK [174], which provides multiple annotated literature data sets to facilitate text mining research in the areas of literature-based database curation, named entity recognition, and protein ontology development.

Swiss-Prot

Swiss-Prot [279] is a protein sequence and knowledge database, and serves as a hub for biomolecular information archived in multiple databases [81]. It is well known for its minimal redundancy, high quality of annotation, use of standardized nomenclature and links to specialized databases. Its format is very similar to that of the EMBL Nucleotide Sequence Database (EMBL-Bank) [81]. Since Swiss-Prot is a protein sequence database, its repository contains amino acid sequences, the protein names and descriptions, taxonomic data and citation information. If additional information is provided with the data, such as protein structures, diseases associated with the protein or splice isoforms, Swiss-Prot provides a table where these data can be stored. Swiss-Prot also combines all information retrieved from the publications reporting new sequence data, review articles and comments from enlisted external experts.

TrEMBL

Due to the large number of sequences generated by different genome projects, the Swiss-Prot database faces several challenges related to the processing time required for manual annotation. For this reason, the European Bioinformatics Institute (EBI), collaborating with Swiss-Prot, introduced another database, TrEMBL (translation of the EMBL Nucleotide Sequence Database), as a supplement to Swiss-Prot. The TrEMBL database consists of computer-annotated entries derived from the translation of all coding sequences in the nucleotide databases. This database is divided into two sections: SP-TrEMBL contains sequences which will eventually be transferred to Swiss-Prot, and REM-TrEMBL contains those which will not go into Swiss-Prot, including patent application sequences, fragments of fewer than 8 amino acids and sequences which have proven not to code for real proteins [114, 134, 279].

UniProt

With protein information spread over multiple data repositories, the efforts from PIR, Swiss-Prot, and TrEMBL were combined to develop the Universal Protein Resource (UniProt) consortium database to centralize protein resources [401]. UniProt is organized into three layers. The UniProt Archive (UniParc) stores the stable, non-redundant, corpus of publicly available protein sequence data. The UniProt Knowledgebase (UniProtKB) consists of accurate protein sequences with functional annotations. Finally, the UniProt Reference Cluster (UniRef) datasets provide non-redundant reference clusters based primarily on UniProtKB. UniProt also offers multiple tools including searches against the individual contributing databases, BLAST and multiple sequence alignment, proteomic tools and bibliographic searches [401].

1.3 PHYLOGENETIC DATABASES

With all of the knowledge accumulated in the genomic and proteomic databases, there is a great need for understanding how all these types of data relate to each other. Since all biological entities have come about through the evolutionary process, the patterns, functions, and processes

TABLE 1.2: Phylogenetic databases and their URLs

Database	URL
TreeBASE	http://www.treebase.org/
TreeFam	http://www.treefam.org/
Tree of Life	http://tolweb.org/tree/
NCBI Taxonomy	http://www.ncbi.nlm.nih.gov/taxonomy
SYSTERS	http://systers.molgen.mpg.de/
PANDIT	http://www.ebi.ac.uk/research/
	goldman/software/pandit/

that they possess are best analyzed in terms of their phylogenetic histories. The same gene can evolve a different timing of its expression, a different tissue where it is expressed, or even gain a whole new function along one phylogenetic branch as compared with another. These changes along a branch affect the biology of all descendant species, thereby leaving phylogenetic patterns everywhere. A detailed mapping between biological data and phylogenetic histories must be accomplished so that the full potential of the data accumulation activities can be realized. Otherwise it will be impossible to understand why certain drugs work in some species but not in others; or how we can design therapies against evolving disease agents such as HIV and influenza.

The need to query data using sets of evolutionarily related taxa, rather than studying single species, has spawned the need to create databases that can serve as repositories of phylogenetic trees, generated by a variety of methods. Phylogeny and phylogenetic trees give a picture of the evolutionary history among species, individuals, or genes. Therefore, there are at least two distinct goals of a phylogenetic database: archival storage and analysis [381]. Table 1.2 lists some major phylogenetic databases and their respective URLs. In what follows we describe each of these databases in more detail.

1.3.1 Phylogenetic Tree Reconstruction Databases

TreeBASE

TreeBASE [318] was developed to help harness the explosively high growth in the number of published phylogenetic trees. It is a relational database and contains phylogenetic trees and the data underlying them. TreeBASE allows the user to search the database according to different keywords and also see graphical representations of the trees. The user can access information such as data matrices, bibliographic citations, taxonomic names, character states, algorithms used and analyses performed. Phylogenetic trees are submitted to TreeBASE by the authors of the journal papers that describe the trees. For data to be accepted by TreeBASE, the corresponding paper must pass the journal's peer review process [318].

TreeFam

TreeFam is a database of phylogenetic trees of animal gene families. Its goal is to develop a curated database that provides accurate information about ortholog and paralog assignments and evolutionary histories of various gene families [120]. To create and curate the trees and families, TreeFam has gathered sequence data from several protein repositories. It contains human (*Homo sapiens*), mouse (*Mus musculus*), rat (*Rattus norvegicus*), chicken (*Gallus gallus*), pufferfish (*Takifugu rubripes*), zebrafish (*Danio rerio*) and fruit fly (*Drosophila melanogaster*) protein sequences retrieved from Ensembl [177], WormBase [75], SGD [76], GeneDB [166] and TIGR [150]. The protein sequences in TreeFam are grouped into families of genes that descended from a single gene in the last common ancestor of all animals or first appeared in animals. From the above sources, families and trees are automatically generated and then manually curated based on expert review. To manage these data, TreeFam is divided into two parts. TreeFam-B consists of the automatically generated trees. It obtains clusters from the PhIGs database [97] and uses BLAST [14], MUSCLE [108], HMMER [105] and neighbor-joining algorithms [315] to generate the trees. TreeFam-A contains the manually curated trees that exploit algorithms similar to the DLI and the SDI algorithm [421]. As more trees are curated, the TreeFam-A database increases while TreeFam-B decreases in size.

1.3.2   Phylogenetic Information Databases

Tree of Life

The Tree of Life (http://tolweb.org) is a phylogenetic repository that aims to provide users with information from a whole-species view. The Tree of Life allows users to search for pages about specific species through conventional keyword search mechanisms. Most interestingly, a user can also navigate through the Tree of Life using hierarchical browsing starting at the root organism, popularly referred to as Life, and traverse the tree until a species of interest is reached. The species Webpage contains information gathered and edited by recognized experts about the species and peer-reviewed resources accessible through hyperlinks.

NCBI Taxonomy Database

The NCBI Taxonomy database [116] is a curated set of names and classifications for all of the organisms represented in GenBank. When new sequences are submitted to GenBank, the submission is checked for new organism names, which are then classified and added to the database. The public taxonomy database, as of 26 September 2011, included 234,991 species with formal names and another 405,546 species without formal name.

There are two main tools for viewing the information in the Taxonomy database: the Taxonomy Browser, and Taxonomy Entrez. Both systems allow searching of the database for names, and both link to the relevant sequence data. However, the Taxonomy Browser provides a hierarchical view of the classification (the best display for most casual users interested in exploring the classification), whereas Entrez Taxonomy provides a uniform indexing, search, and retrieval engine with a common mechanism for linking the taxonomy and other relevant Entrez databases.

1.3.3 Protein Databases with Phylogenetic Information

SYSTERS

SYSTERS is a protein clustering database based on sequence similarity [257]. SYSTERS (release 4) contains 969,579 non-redundant sequences (and annotations of 1,168,542 redundant sequences totally) gathered from existing sequence repositories: Swiss-Prot [279], TrEMBL [279] and complete genomes, Ensembl [177], *Arabidopsis* Information Resource [258], SGD [76] and GeneDB [166], which are sorted into 158,153 disjoint clusters. Two innovative features of this repository are the SYSTERS Table and SYSTERS Tree. The SYSTERS Table for a family cluster contains a variety of information, most notably accession numbers as well as accession numbers for a variety of external databases including IMB [306], MSD [49], ENZYME [29], INTERPRO [269], PROSITE [179] and GO [26]. There can be several redundant entries in the table for one protein sequence. Since SYSTERS data relies on external protein databases, it always contains an entry name (protein name) and an accession number for each entry but there may not be a gene name. For each family cluster that consists of more than two non-redundant entries, a phylogenetic tree is available. The phylogenetic trees are constructed using the UPGMA method [103].

PANDIT

PANDIT (Protein and Associated Nucleotide Domains with Inferred Trees) is a non-redundant repository of multiple sequence alignments and phylogenetic trees. The database consists of three portions: protein domain sequence alignments from Pfam Database [34], alignments of nucleotide sequences derived from the EMBL Nucleotide Sequence Database (EMBL-Bank) [81], and phylogenetic trees inferred from each alignment.

PANDIT (Version 17.0) contains 7738 families of homologous protein sequences with corresponding DNA sequences and phylogenetic trees. All alignments are based on Pfam-A [343] seed alignments, which are manually curated and, therefore, make PANDIT data high quality and comparable with alignments used to study evolution. Each family contains three alignments: PANDIT-aa containing the exact Pfam-A seed protein sequence alignment; PANDIT-dna containing the DNA sequences encoding the protein sequences in PANDIT-aa that could be recovered; and PANDIT-aa-restricted containing only those protein sequences for which a DNA sequence has been recovered. The DNA sequences have been retrieved using cross-references to the EMBL Nucleotide Sequence Database (EMBL-Bank) from the Swiss-Prot [279] and TrEMBL [279] databases. To ensure accuracy, PANDIT performs a translation of the cross-referenced DNA sequences back to the corresponding protein sequences.

The PANDIT database is intended for studying the molecular evolution of protein families. Therefore, phylogenetic trees have been constructed for families of more than two sequences. For each family, five methods for tree estimation have been used to produce candidate trees. These methods include Neighbor-joining [315], BioNJ [133], Weighbor [58], FastME [99] and PHYML [145]. Neighbor-joining, BioNJ and Weighbor are used to produce phylogenetic tree estimates from a pairwise distance matrix. FastME uses a minimum evolution criterion with local tree-rearrangements to estimate a tree, and PHYML uses maximum likelihood with local tree searching. At the end, the likelihood of each tree from the candidate set is computed and the tree with the highest likelihood is added to the database.

1.4 STRUCTURE AND PATHWAY DATABASES

Knowledge of protein structures and molecular interactions is key to understanding protein functions and complex regulatory mechanisms underlying many biological processes. However, computationally, these data sets are highly complex. The most popular ways to model these data sets are through text, graphs or images. Text data tends not to have the descriptive power needed to fully model these types of data. Graphical and image data require complex algorithms that are computationally expensive and not reliably accurate. Therefore, structural and pathway databases become an interesting niche from both the biological and computational perspectives. Table 1.3 lists several prominent databases in this field and their respective URLs. In what follows we describe these databases in more detail.

1.4.1 Protein and Nucleic Acid Structure Databases

Protein Data Bank

The Protein Data Bank (PDB) is an archive of structural data of biological macromolecules. The PDB is maintained by the Research Collaboratory for Structural Bioinformatics (RCSB). It allows the user to view data both in plain text and through a molecular viewer utilizing Jmol. A key goal of the PDB is to make the data as uniform as possible while improving accessibility and providing advanced querying options [40]. In order to have complete information regarding the features of macromolecular structures, the PDB allows a wide spectrum of queries through data integration. The PDB collects and integrates external data from scientists' deposition, Gene Ontology (GO) [85], Enzyme Commission (http://www.chem.qmul.ac.uk/iubmb/enzyme/), KEGG Pathway Database [191], and NCBI resources [7]. The PDB realizes data integration through data loaders written in Java, which extract information

TABLE 1.3: Structure and pathway databases and their URLs

Database	URL
Protein Data Bank	http://www.rcsb.org/pdb/
Nucleic Acid Database	http://ndbserver.rutgers.edu/
KEGG	http://www.genome.jp/kegg/
BioCyc Database Collection	http://www.biocyc.org/

from existing databases based on common identification numbers. The PDB also allows data extraction at query run time, which means implemented Web services extract information as the query is executing.

Nucleic Acid Database

The Nucleic Acid Database (NDB), also curated by RCSB and similar to the PDB and the Cambridge Structural Database [12], is a repository for nucleic acid structures. It gives users access to tools for extracting information from nucleic acid structures and distributes data and software. The data are stored in a relational database that contains tables of primary and derivative information. The primary information includes atomic coordinates, bibliographic references, crystal data, data collection and other structural descriptions. The derivative information is calculated from the primary information and includes chemical bond lengths and angles, virtual bond lengths and other measures according to various algorithms [28, 144]. The experimental data in the NDB database have been collected from published literature, as well as from one of the standard crystallographic archive file types [144, 222] and other sources. Primary information has been encoded in ASCII format (American Standard Code for Information Interchange, ASA X3.4-1963, American Standards Association, June 17, 1963). Several programs have been developed to convert among various file formats [272].

1.4.2 Functional Pathway Databases

KEGG

The Kyoto Encyclopedia of Genes and Genomes (KEGG) [191] is the primary resource for the Japanese GenomeNet service that attempts to define the relationships between the functional meanings and utilities of cell or organism and its genome information. KEGG contains three databases: PATHWAY, GENES and LIGAND. The PATHWAY database stores computerized knowledge on molecular interaction networks. The GENES database contains data concerning sequences of genes and proteins generated by the genome projects. The LIGAND database holds information about the chemical compounds and chemical reactions that are relevant to cellular processes. KEGG computerizes the data and knowledge as graph information. The PATHWAY database contains reference diagrams for molecular pathways and complexes involving various cellular processes that can readily be integrated with genomic information [192]. It stores data objects, called generalized protein interaction networks [189, 190]. The PATHWAY database is composed of four levels that can be accessed through a Web browser. The top two levels contain information about metabolism, genetic information processing, environmental information processing, cellular processes and human diseases. The others relate to the pathway diagram and the ortholog group table, which is a collection of genes and proteins.

BioCyc Database Collection

The BioCyc Database Collection [196] is a compilation of pathway and genome information for different organisms. Based on the number of reviews and updates, BioCyc databases are organized into several tiers. As of September, 2012, Tier 1 consisted of six intensively curated databases. Tier 2 contained 42 databases computationally generated by the PathoLogic program. These databases have been updated and manually curated to varying degrees. Tier 3 is composed of 1920 databases computationally generated by the PathoLogic program with no review or updating [196].

The BioCyc Website allows scientists to perform certain operations, e.g., visualize individual metabolic pathways, view the complete metabolic map of an organism, and analyze metabolomics data using the Omics Viewer. The Website also provides a spectrum of browsing capabilities such as moving from a display of an enzyme to a display of a reaction that the enzyme catalyzes or to the gene that encodes the enzyme [196].

1.5 MICROARRAY AND BOUTIQUE DATABASES

Both microarray databases and boutique databases offer interesting perspectives on biological data. The microarray databases allow users to retrieve and interact with data from microarray experiments. Boutique databases offer users specialty services concerning a particular aspect of biological data. Table 1.4 lists some of these databases and their respective URLs. In what follows we describe each of these databases in more detail.

Gene Expression Omnibus

The Gene Expression Omnibus (GEO) [32, 107] at the National Center for Biotechnology Information (NCBI) is the leading public repository for gene expression data. Gene expression is the process of synthesizing a product from a DNA gene. A gene product is often a protein, but in a non-protein coding gene, the product may be one of a wide variety of functional RNAs, also known as non-coding RNAs (ncRNAs). Gene expression data are analyzed to identify how an organism functions in a normal situation or a stressful situation, such as disease or starvation. Much can be learned about how a cell functions by comparing the gene expression data from a diseased cell to data from a healthy cell. As of 2012, GEO contained over 800,000 samples derived from over 1,600 organisms [33]. The source of most of the gene expression data in GEO was microarray profiling. However, since 2008, next-generation sequencing (NGS) submissions have been rapidly increasing. GEO datasets may be browsed at http://www.ncbi.nlm.nih.gov/sites/GDSbrowser [32]. A Web application named GEO2R that provides R-based analysis of GEO data is available at http://www.ncbi.nlm.nih.gov/geo/geo2r/ [33].

RegulonDB

RegulonDB is a manually curated database of *Escherichia coli* K-12 transcriptional regulation. RegulonDB serves as an authoritative reference for *E. coli* regulons, which currently number around 500 [316]. A regulon is a group of genes regulated as a unit by a common transcription factor or regulatory gene [236]. When the *E. coli* genome was first sequenced in 1997, just 99 of its regulons were known. [44]. RegulonDB is a relational database providing the bacteria research community with organized and computable knowledge on *E. coli* transcriptional regulation.

Yale Microarray Database

The Yale Microarray Database (YMD) [77] is another repository for gene expression data. It is Web-accessible and enables users to perform several operations, e.g., tracking DNA samples between source plates and

TABLE 1.4: Microarray and boutique databases and their URLs

Database	URL
Gene Expression Omnibus (GEO)	http://www.ncbi.nlm.nih.gov/geo/
RegulonDB	http://regulondb.ccg.unam.mx/
Yale Microarray Database	https://medicine.yale.edu/keck/ymd/
ESCAPE Stem Cell Database	http://www.maayanlab.net/ESCAPE/
Neurodatabase DataServer	http://www.neurodatabase.org

arrays, and finding common genes or clones across different microarray platforms. Moreover, it allows the user to access the image file server, enter data, and get integrated data through linkage of gene expression data to annotation databases for functional analysis [77]. YMD provides several means of querying the database. The Website contains a query criteria interface [77] that allows the user to perform common queries. The interface also enables the user to choose the format of the output, e.g., columns to be included and the type of output display (HTML, EXCEL, TEXT or CLUSTER). Finally, the query output can also be dynamically linked to external annotation databases such as DRAGON [48].

ESCAPE Stem Cell Database

The ESCAPE Stem Cell Database supported by Mount Sinai School of Medicine (MSSM) is a unique repository that contains information about stem cells from mouse and human species [406]. The Embryonic Stem Cell Atlas from Pluripotency Evidence (ESCAPE) is mammalian embryonic stem cell (ESC)-specific. Data in this repository are created by collecting and integrating results from various published studies that profiled human and mouse ESCs. Users can query various known and novel regulatory interactions across various regulatory layers.

Neurodatabase DataServer

The Neurodatabase DataServer supported by Cornell University is a repository containing data that pertain to the understanding of neural coding, information transmission and brain processes, and provides a venue for sharing neuro-physiological data. It acquires, organizes, annotates, archives, delivers, and displays single- and multi-unit neuronal data from mammalian cerebral cortices [129, 130]. Users can obtain the actual datasets provided by several laboratories, all in common format, and employ BrainML, a data-description language and functional ontology. The Web interface provides a feature called Query Tool that allows the user to search by metadata terms submitted by the researchers. A Java tool known as the Virtual Oscilloscope displays time-series and histogram datasets dynamically. The datasets can also be downloaded for analysis.

CHAPTER 2

Biological Data Cleaning

CONTENTS

2.1   INTRODUCTION

Bioinformatics databases, like most databases, are subject to many problems. Bioinformatics databases contain data that do not conform to the original intent of the databases, or "dirty" data. However, due to the nature of biological data, usually these problems become complex, thereby excluding the more traditional methods for solving "dirty" data problems. Conceptually, a framework for cleaning and integrating biological data would need to address both the schema and the data issues that cause data quality problems within a bioinformatics database [302].

2.1.1   Data Quality

Schema-related problems, usually fall into a few categories. First, a bioinformatics database is part of a federated database system or interacts with other databases to obtain data. These data may or may not use the same schema used by the database. Therefore, these data need to be mapped onto the database's schema.

     This is a mapping of multiple autonomous database systems into a single "virtual" database. The multiple autonomous database systems may belong to different organizations that may be geographically decentralized. A federated database contains none of its own data (hence the "virtual" reference), but rather is a composite of the autonomous database systems that comprise it. The federated database system provides a uniform user interface to multiple similar systems that typically have disparate schemas. A federated database is sometimes a feasible alternative to merging multiple databases into a single system, which may be a fairly complex process.

The second category of schema-related problems occurs when an error is made in the design of the database's schema. This can very easily happen in bioinformatics databases since most schemas need to model complex biological and chemical structures as well as the metadata involved with the data. A method may be needed by a bioinformatics database to modify the schema of the database while preserving the data.

Mapping the new schema to the old schema can help to solve this problem. Also, mapping can be viewed as a data integration problem. Since all data submitted to the repository needs to be included with the current dataset, the submission problem can be seen as an integration issue. Moreover, if the repository transfers to a new schema, the legacy data needs to be integrated to interact properly with the data in the new schema. In addition, if the repository interacts with other repositories, the result is a classical data integration problem.

Since many of the biological fields are dynamic, with new knowledge generated very quickly, schemas can change or be inadequate to handle new data. If a database administrator changes schemas to create a more powerful database, issues of integrating the legacy schema and legacy data arise. Any data that conforms to the legacy schema may be missing data expected by the new schema. The legacy data may not conform to the new schema. Therefore, the legacy data may need to be mapped to the new schema [302].

Besides the schema-related problems, there are other data obstacles. First, data can be entered into the database erroneously [302] and will need to be corrected. Due to the dynamic nature of bioinformatics databases, and since these databases can house anything from protein structures to strings about the genetic composition of a species, it is highly possible that data duplication occurs. This duplicate data will need to be identified and properly organized.

Secondly, the legacy data will need to be integrated properly into the database. This will require mapping it properly as well as using various detection mechanisms to discover whether this data previously existed. Thirdly, much of the data within bioinformatics databases is generated through the use of computational tools. Errors can occur through improper implementation of an algorithm or improper execution of a tool. For example, researchers use reconstruction algorithms applied to taxa data to generate phylogenetic trees. If the reconstruction algorithm is instantiated improperly, the data will be skewed toward that error. This can have serious consequences since, if the data is used for further purposes, the error will be propagated throughout any application using that data.

Over the past several years, the proliferation of bioinformatics databases, data banks and data warehouses has been remarkable. There are large databanks that act as repositories for genetic data and their mirror sites. Examples are GenBank [37] and its partners, EMBL-Bank (Europe) [81] and the DNA Data Bank of Japan (DDBJ) [280] discussed in Chapter 1. There are also specialized or "boutique" databases that specialize in specific functionalities, as surveyed in Chapter 1. These databases can also be curated in a number of ways. Most are public repositories, where data can be submitted through the World Wide Web. Examples of this type of database are GenBank and the Protein Data Bank (PDB) [38]. GenBank allows automated submissions but depends on a curator review before submission is complete. The PDB, with the ADIT tool [390, 391], has a fully automated annotation and deposition process. However, there are still instances where a curator needs to intervene. These databases have created new and interesting problems for database researchers.

Bioinformatics databases tend to have specific needs that are not normally addressed with common databases and tend to be dynamic. Data is constantly added to these databases. This process of adding data can take place through a number of methods. Some databases use a manual editing approach while others allow authors to submit data through World Wide Web forms. While most manually edited data are still submitted through a Web interface, a curator needs to review the contents of a submission. The curator then can manipulate or annotate the data according to the need of the database. Examples of this type of database include GenBank and TreeBASE. As mentioned above, the Protein Data Bank uses a fully automated tool called ADIT for deposition. ADIT allows a researcher to submit data through the World Wide Web. ADIT proceeds to process the submission and ensures it conforms to the standard the Protein Data Bank uses. This submission can also be annotated manually as needed [390, 391]. Moreover, knowledge related to this data is also constantly being discovered. The knowledge can affect everything from the data to the schema. It also affects legacy data within the database. It can have consequences with analysis of the data and interoperability among the databases. Looking at phylogenetic data as a model for applying data cleaning to biological data, a number of issues are both interesting and unique to the phylogenetic field.

Phylogenetic data relate to the evolutionary knowledge of a particular species. Phylogenetics is an area of study dedicated to researching the evolutionary history and relationships of species. The primary goal of phylogenetics is to understand how species evolved. Understanding the evolution of an organism or species is a complex task that requires biologists to understand both the molecular composition of a species and the effects of environment on its development. The molecular composition of a species describes its genetic structure as well as the arrangement of its biological system. The effect of the species' environment considers all outside factors that affect its development and can include information about its place in the food chain, its mating practices, host-parasite relationship with other species as well as many other factors. These environmental effects as well as biological composition determine the possibility for the perpetuation of the species [286]. Moreover, through trying to understand the evolutionary history of a species, biologists can uncover its relationship to other species. It is theorized that all life comes from one species. Phylogenetic research allows biologists to try to trace the development of current species to this single original species by organizing phylogenetic trees.

Phylogenetic trees are structures that model the evolutionary relationships among species. A phylogenetic tree's leaf nodes represent existing species. Internal nodes represent species from which the species in the leaf nodes descend. Phylogenetic trees are based on the various characteristics of species. Currently, the most popular method of developing phylogenetic trees, called phylogenetic reconstruction, compares the genetic structures of a set of species and decides how closely they are related. The species most closely related become siblings or closely related through a common ancestor. Usually a node will have a descendent if a significant change occurs within the genetic structure of the species the node represents, such that a new species or a set of species results. However, there are other phylogenetic reconstruction methods that consider different factors when building phylogenetic trees [286].

Understanding phylogenetic information about a species can be very beneficial in many other areas of research. By understanding the phylogenetic information about a species, we can understand the classification of the species. For example, understanding the evolutionary development of whales can explain why whales are classified as mammals instead of fish. Furthermore, it can help define the classification system for species, thereby giving scientists criteria for classifying previously unknown species. It can help predict the future evolutions of the species or explain why two species with a common ancestor branched from that ancestor and formed new species. It can also help track the evolution of a species. For example, the mutation of viruses can be tracked this way. Since viruses have distinct genetic compositions, and work through mutations, various strains of the same virus can be analyzed using phylogenetic analysis of their genetic composition. Phylogenetic trees can be created from this information, possibly showing which strain of the virus is older. This could then give scientists indications as to how the virus was transmitted from one host to another, showing contagion patterns [286].

With the developmental information about the species as well as the information generated by the phylogenetic reconstructions, the data surrounding phylogenetic studies become complex and difficult to manage. These complex data need various types of tools that will allow users to effectively manipulate the data. It becomes imperative that these data are as "clean" as possible. Issues concerning cleaning a phylogenetic database become complex since the database works with data in traditional record formats and also the database contains data that model the structures of the phylogenetic relationships.

Bioinformatics databases need to interact with other databases. A database needs to be clean so that integration with new submission data and existing data from other sources is possible. Issues develop about how to clean text, since text can represent biological sequences or names of species. Data of primary interest are components of phylogenetic trees. The issues concerning how to clean phylogenetic trees are important. For example, there is a need to ensure that data for every species that is a member of a phylogenetic tree is correct since errors can create input errors. The issue of whether this tree exists already within the database also becomes of interest and can reveal a duplicate record or related information a user might be interested in. Knowing what trees are similar within the database is extremely important. This can also indicate duplicate records. If other aspects of the phylogenetic data are included within this similarity test, informative comparisons can be made about the algorithms used to create the trees. Testing can indicate duplications or errors in research. For example, if two researchers use the same method to develop similar trees yet the trees are very different, their results indicate an error in submission of the trees, an error in the tools the researcher used or an error within the algorithm.

When analyzing bioinformatics databases generally, a number of data problems can be identified. First, most data in these databases have been accumulated. Commonly, no strict database rules have been applied to it, causing erratic schemas. These schemas can be difficult to understand by visual inspection and even more difficult to design automated tools around for knowledge discovery. As many of these databases age, data uniformity becomes an issue. In the biological fields, innovation and knowledge about the biological systems studied grow every day, adding to the knowledge about the data. More aspects of the data stored today are known than were known as recently as 5 years ago. This causes inconsistencies within the data format. The Protein Data Bank (PDB) illustrates this issue very well. PDB staff dedicate significant resources to their "data uniformity project" [390, 391]. The data within the PDB are representative of over 30 years of research within microbiology and protein chemistry. Knowledge has increased considerably during this period, causing the amount and type of information stored to change. For example, knowledge about the atomic structure of proteins has changed and required that the data schema respond to such information changes. However, previous data may have been maintained in obsolete formats. Also, there are nomenclature issues, for example, various atoms are misnamed. The old formats need to be reconciled along with nomenclature issues to make the database uniform [390, 391]. Besides causing data inconsistencies, the greater issue is how to update this data. Also, since the new data have more aspects, the schema tend to be different from earlier data.

Such "dirty data" occurrences cause other issues. For example, since there are multiple formats, and biological data are extremely complex, the problem of synonymy arises. In biological research, it is very easy for researchers to consider a species, disease or protein to be new while in reality it is only a minor mutation of an already discovered biological entity. However, since the data sets are so large and the data management so difficult, it is difficult to detect these problems.

Two examples of noisy data problems can be seen in the phylogenetic tree database TreeBASE [284]. TreeBASE, which makes its data public through the matrix files for their drawing tools, has a number of inconsistencies. TreeBASE, like many other bioinformatics databases, hosts many files that contain legacy data. As knowledge about the data increased, formats changed. For example, the representations for the genetic codes differ from file to file, usually depending upon when the data was researched. Representation of gaps within the genetic code is also file dependent and representations for the trees themselves change slightly from file to file. While all trees are in Newick notation [67], the metadata surrounding the trees change format. Therefore, it can be difficult to process the trees.

The issues of data integration and interoperability are also important. As more and more bioinformatics databases develop, the need to be able to integrate information seamlessly from one tool to another becomes important. Even among data banks where the data stored is nearly identical, many databases have specialized tools that no other

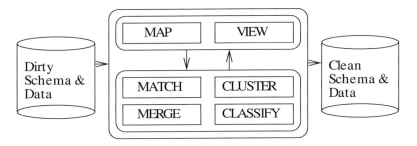

FIGURE 2.1: A framework for biological data cleaning.

databank has. Ideally, it would be advantageous to allow biologists to use whatever tools they want to analyze whatever data they want.

2.1.2 Framework

Figure 2.1 presents a framework for biological data cleaning. This framework should ultimately reduce if not eliminate "dirty" data and schemas from the database. It should return to the user a clean database. To do this, first the bioinformatics database must be analyzed. The issues concerning the schema and those concerning the data need to be separated. Next, various cleaning operations will be performed upon the schema and data. These cleaning operations include a method for modifying schema (MAP), detecting duplicates or errors (MATCH, CLUSTER, MERGE and CLASSIFY), and viewing any section of the database needed (VIEW) [125]. Once the data and schema are cleaned, the new data is staged for re-integration into the database. This allows integration of the cleaned data to be performed as well as any other operations needed before the data is re-integrated. Finally, the data and schema are re-integrated, resulting in a cleaned database.

While this framework is conceptual, there are a few requirements needed so that the framework can perform effectively on bioinformatics databases. Primarily, it must be extensible. As will be further explored in the general data cleaning section, many cleaning frameworks are not extensible. If the framework is not extensible, it cannot be flexible enough to support the various operations needed to clean biological data. Biological data tend to be complex, incorporating structural data, keyword-based data, metadata and other forms of data. With the multitude of bioinformatics databases available on the Web as surveyed in Chapter 1, a framework must be able to consider the different types and different categories of data these databases can hold. An extensible framework is imperative to respond to these needs.

2.2 GENERAL DATA CLEANING

As mentioned above, data cleaning methods address the issues concerning the quality of data [91, 92]. It is also sometimes referred to as the process of taking data that is "noisy," sometimes also referred to as "dirty," or otherwise in a format that computational tools have difficulty processing and transforming into a more accessible format. This process maintains the integrity of the data while removing aspects that can be considered "dirty" [91, 92].

In data cleaning, three categories of problems need to be dealt with. The first is concerned with erroneous data detection. Problems from erroneous data usually stem from, but are not limited to, user input errors, inconsistency in input, missing values, misspellings, improper generation of data, and legacy data differences [91, 92].

The second category is concerned with duplicate detection. At first, duplicate detection was applied to very large databases where control was not very strong. The first algorithms were essentially string detection and record detection methods for relational databases. However, as data and schemas grew more complex, duplication detection became harder. Also, whether two similar documents were actually duplicated became a pressing question. With simple similarity detection through sorting and joining records within a database, more complex duplication errors became evident. Duplication can occur and not be detected through data errors. Also, as data becomes more complex, synonymy can occur when two records are not identical syntactically, but are identical semantically [302].

The third category concerns schema errors and schema and data integration. Schema errors result from improperly formed schemas or schemas that need to change frequently. Schema and data integration occurs when the database is part of a federated system or needs to be synchronized with other databases. With interoperability facilitated through the World Wide Web, many databases tend to need to synchronize with other databases. This synchronization includes communicating both data and schemas between databases and updating the databases accordingly. Therefore, a proper schema becomes essential for communicating with other databases. Moreover, for rapidly changing data, schemas sometimes need to be modified to reflect these changes. Without properly mapping previous data, the database can have trouble arising from legacy data interacting with new data [302].

Data cleaning problems have been observed throughout the history of databases, with the first documented cases stemming from the United States Census information during the 1950's. Since then, there have been many developments within the data cleaning field. Commonly, a data cleaning system should fulfill certain conditions. First, it should remove errors within the data as well as schema integration issues for federated databases. Second, the cleaning should be done through automated tools rather than manual inspection. Third, data cleaning should not be done in isolation but rather with respect to the entire database and its schema [302].

2.2.1 Duplicate Elimination

One of the earliest methods for cleaning data was duplicate detection and elimination. One of the most popular methods for duplicate detection was introduced by Bitton, et al. [41]. At the time, for most database management systems, trivial duplicates could be found through sorting and then performing a join procedure that would find the duplicates and they could be eliminated. However, this was a time-intensive procedure. Many database management systems chose not to sort in favor of speed. Bitton's method was to modify the sort procedure so that it ran somewhat faster than the sort procedure provided by the relational databases of the time. Also, this procedure, upon detecting a duplicate, immediately removed the duplicate, thereby eliminating the need for the join procedure. The sorting algorithm can be any efficient sorting algorithm that allows for external sorting. Some of these algorithms include two-way merge sorts and the use of hash tables [41].

Since Bitton's method was introduced, extensive work has been done to detect duplicates and approximate duplicates within databases. One of the major advancements in this approach to duplicate detection is the sorted neighborhood technique. In the sorted neighborhood approach, a key is extracted from the data. This key, as a unique identifier for the data, also represents the minimum set of elements from which similarity can be determined. This key is usually a combination of the elements within the given set of attributes rather than all of the data within the attributes. This increases the speed of the sorting as well, and gives a smaller string for comparison. Records are then sorted based on this key. From this sorted list of records, similarity can be measured through the use of a "window." This window W represents W records from the sorted set. Any record R within W is then compared with the $W - 1$ records before R and $W - 1$ records after R in sorted order. If the records then are judged to be similar, the duplicates are eliminated [163, 164, 240].

While the sorted neighborhood technique detects many duplicates, it does have limitations. For example, it detects duplicates only within the given window size. Moreover, duplicate detection is highly dependent upon the key selected for sorting the records. Based on these observations, Hernandez, et al. [163, 164] developed a modified sorted neighborhood technique. They perform the sorting through multiple iterations. During each sort, the key is modified so that the sort results will be different. After each sort, the records are compared, with duplicates deleted and similar records merged. Since there are multiple passes through this sort-and-merge algorithm, more records that are similar can be detected. Also, relating records that are not obviously similar becomes easier by using the union of the sets of results obtained from each sort and obtaining the transitive closure of the union [163, 164].

There is also an incremental method to cleaning databases based on this multiple-sort sorted neighborhood technique. In this method, the objective is to not clean previously cleaned data, but to clean new data within the database and adjust the records about the previously cleaned data to reflect this new cleaning. In the incremental sorted neighborhood algorithm, new data is compared against keys that are representatives of the database. These representative keys are obtained from the previous cleanings. In each of the previous cleanings, similar records are grouped together or clustered and from these clusters, a key describing the common attributes from the cluster is developed. When performing a new cleaning, a dirty record is compared against the representatives through the multi-pass version of the sorted neighborhood method. If duplicates are detected during this phase, they are merged or eliminated. Once the various sorts finish, based on the knowledge gained from the sorts, the record is placed into the cluster to which it is most similar. The representative of the cluster is then modified to reflect the new record [163, 164].

2.2.2 Knowledge-based Methods

Besides using methods with various applications of sorting, there are a number of machine learning methods for performing data cleaning [80, 240]. One popular method is to incorporate knowledge bases into data cleaning tools. Knowledge bases provide domain-dependent information that can be used to improve error detection techniques and detect duplicates. A good example of a knowledge-based data cleaning tool is Intelliclean by Lup Low, et al. [240]. Intelliclean first "scrubs" the data for irregularities that can be detected easily. For example, it standardizes abbreviations. Therefore, if one record abbreviates "street" as "St." and another abbreviates it as "Str." while another record uses the full word, all three records can be standardized to the same abbreviation. Once the data have been scrubbed, they are cleaned using a set of domain-specific rules that work with a knowledge base. These rules detect duplicates, merge appropriate records and create various alerts for other anomalies.

Finally, the cleaning is validated through user interaction and inspection [240].

While many data cleaning tools are automated, other tools require user interaction to perform cleaning. Potter's Wheel developed by Raman and Hellerstein [305] at the University of California at Berkeley is such a system. With interactive tools, data cleaning can be more focused toward the user's specific needs. Potter's Wheel is a domain-independent tool. The interface allows the user to view the various cleaning rules employed on the data set. From this view, the user can then modify the rules so that cleaning becomes more precise. The actual operation to clean the data is done without the user interaction with the system [305].

Another method by Caruso, et al. [68] uses data reconciliation techniques to match and combine duplicate records within a database. They use an extensible platform that uses machine learning techniques to decide how to eliminate duplicate entries within a database. The tool uses a training set of data from a database to which it will eventually be applied. From this training set, the tool can develop a set of rules for matching data to find duplications. Once the training set is selected, pre-processing of the data occurs. This pre-processing can include removing common words within the data set or reducing white space. Next, a set of measures of similarity is selected. The tool is then trained on the selected data to obtain the rules for detecting duplication and then applied to the entire data set [68].

2.2.3 Extraction, Translation and Loading Method

One of the most popular methods for data cleaning is the extraction, translation and loading method (ETL). During this process, two types of cleaning occur: on the instance level and on the schema level. Instance level cleaning refers to errors within the data such as misspellings. Schema level cleaning usually concerns integrating the database into a new schema, a data warehouse or a federated database. In the ETL process, the data and schema are extracted, various operations are performed on the data and schema to clean them, and the new schema is put into the database with the cleaned data [302]. ETL's primary tools are data flow graphs that track the transformation of the dirty data into the cleaned data [302]. The ARKTOS system, a data cleaning system that implements the ETL method, is an example of a system that uses graphs [372].

While each of aforementioned methods is interesting, they each have drawbacks. The ETL method tends to be database-specific. The tools designed to perform the extraction, translation and loading tend to apply to only one database. The tool by Caruso, et al. only considers the need for matching and deleting duplicates. Also, the pre-processing phase at which a set of data is selected for training has some disadvantages. First, it requires prior knowledge of what instances in the database might be problematic. Second, the data also need to be standardized before the learning algorithms are applied. A promising method that does not have these drawbacks is the declarative method in which various operators for data cleaning are conceptually defined. Then, tools are developed to implement these operators, depending upon the definition of the concept with respect to the data. For example, a match in a dictionary would be defined differently from a match for a protein in a bioinformatics database. This method has been implemented into the AJAX tool.

AJAX, developed by Galhardas, et al. [125], is a data cleaning system for publication and citation information. The cleaning operations are applied to this data, which is in a relational database format. These operations present an approach to perform cleaning on most data sets. Since they are not instantiated precisely, but rather can be instantiated into whatever data cleaning system a group wants to develop, they can be instantiated based on the data set. This preserves the concepts of the operations needed to perform data cleaning while also allowing flexibility a specific data set needs from a particular operation. Therefore, these data cleaning operations can be applied to biological and evolutionary data. Moreover, the operations can also be applied regardless of the type of data model or schema used to create the database.

2.2.4 Declarative Data Cleaning

Table 2.1 includes the syntax for the mapping operator from [125]. Also included are the syntaxes for the *let* clause (Table 2.2) and the *output* clause (Table 2.3), which are part of the syntax for the mapping clause as well as some of the operators introduced in this section.

TABLE 2.1: AJAX *mapping* operator syntax

\<mapping-operator\>	:	create mapping \<operation-name\>
		from \<predicate-name\>
		[\<alias-variable\>]
		[let \<let-clause\>]
		[where \<where-clause\>]
		\<output-clause\>

The syntax for the mapping clause is very similar to that of an SQL statement. The "create mapping <operation-name>" line specifies what operation is to be performed as well as an identifier for that operation. The "from" clause specifies the standard SQL FROM clause. It sets the proper input predicate as well as the alias variable associated with the input predicate to be used throughout the rest of the operator statement [125].

TABLE 2.2: AJAX *let* clause syntax

<let-clause>	:	<assignment-statement>, [{<assignment-statement>} ...]
<assignment-statement>	:	<predicate-name> = <functional-assignment>
<functional-assignment>	:	<functional-expression> | <if-then-else-expression> | <sfw-expression>
<if-then-else-expression>	:	if <condition> then <then-else-expression> [else <then-else-expression>]
<then-else-expression>	:	throw <exception-name> | <functional-assignment>
<functional-expression>	:	<function-name> (<arg-expression> [{,<arg-expression>} ...])
<arg-expression>	:	<constant> | <domain-variable> | <predicate-name> | <functional-expression>
<sfw-expression>	:	select <sql-project-clause> from <sql-from-clause> where <sql-where-clause>

The "where" clause is similar to the SQL WHERE clause, in which it acts as a filter for the operation. The "let" clause as used in a few of the AJAX operations creates a predicate key for the operation by calling to atomic functions that generate the key. The let clause's syntax in Table 2.2 includes functional assignments (<functional-assignment>), control statements (<if-then-else-expression>), SQL-like statements (<sfw-expression>), and exception handling (throw <exception-name>). This gives the mapping powerful control over the data. The functional assignments are atomic statements that allow a predicate to be assigned a value. The control statements allow for selection of specific statements and exception handling. Finally, the SQL-like statements allow for further refinement of the data [125].

TABLE 2.3: AJAX *output* clause syntax

<output-clause>	:	<select-into-clause> [{<select-into-clause>} ...]
<select-into-clause>	:	<sql-select-into> [{constraint <constraint-clause>} ...]
<sql-select-into>	:	select <sql-project-clause> into <predicate-name>
<constraint-clause>	:	unique <att-name>[{<att-name>} ...] | not null <att-name> [{<att-name>} ...] | foreign key <att-name> [{<att-name>} ...] references <predicate-name> (<att-name> [{<att-name>} ...]) | check <where-clause>

The "output" clause organizes the data wanted from the mapping operation (or any other operation using the clause) into the desired schema. The output clause, whose syntax is shown in Table 2.3, is a modified SQL statement. The "select-into" statement stipulates the schema of the target relation as well as the constraints on the relation [125].

To address some of the problems identified in bioinformatics databases as well as those raised concerning some of the data cleaning methods, the declarative data cleaning method can be used. The method developed by Galhardas, et al. in "Declarative Data Cleaning: Language, Model and Algorithms" [125] defines a method, or more precisely a set of operations essential to any data cleaning tool. These operations utilize a BNF-styled notation where non-terminal symbols are contained within angled brackets '<' and '>'; alternative productions are indicated with

'|'; terminal symbols are indicated in capital letters, a production enclosed within brackets '[ ]' indicates that production is optional and a production enclosed within curly braces '{ }' can be repeated one or more times [125].

Mapping Operator

Mapping generally is defined as an operation that takes a tuple from one relation and transforms it into a tuple for another relation. For example, given two databases for personal information, the name of a person could be represented as (lastName, firstName, middleInitial) while in a second database, this information could be represented in one as (lName, fName, mIn). A mapping operator would specify how to transform the first schema into the second. The mapping operator can be very useful for the creation of new relations as well as transforming data from alternative sources into a format the database can process [125].

View Operator

The view operator corresponds to an SQL query with modifications that can help with exception handling and integrity constraints [125]. The syntax of the view operator shown in Table 2.4 is the syntax of any SQL statement that can be used to extract information from the database. The "create view" statement creates the predicate that will hold the information obtained from the SQL statement in the <operation-name> predicate [125].

TABLE 2.4: AJAX *view* operator syntax

<view-operator>	:	create view <operation-name>
		from <predicate-name> [<alias-variable>]
		[{<predicate-name> <alias-variable> ... }]
		[where <where-clause>]
		<select-into-clause>

Matching Operator

The matching operator computes the distance between two input tuples to determine the degree of similarity between the tuples. This function can be used to detect duplicates within the database and detect errors against a knowledge base. Also, it can be modified so that the degree of similarity can be computed. In a traditional database, matching can be computed through joining two relations and using an arbitrary distance measure [125]. The syntax of the matching operator is given in Table 2.5.

The matching operator's syntax is very similar to those of the mapping operator and the view operator. The matching operator's syntax is primarily that of an SQL statement. However, to perform the more advanced matching operations, the let clause is used. Also, the [+] in the from clause indicates that the predicates that do not have a match should be returned as well [125].

TABLE 2.5: AJAX *matching* operator syntax

<matching-operator>	:	create matching <operation-name> from <predicate-name> [+] <alias-variable> [{<predicate-name> [+] <alias-variable> ... }] [let <let-clause>] [where <where-clause>] into <predicate-name>

Cluster Operator

The cluster operator takes a relation that defines a set of elements from the database as input and returns one relation as output. This operation allows for the definition of that output relation to be a description of some distance within the initial set of elements. Generally, there are two methods from which clustering can be performed. Clustering can be performed based on the data value within a specific attribute. This is similar to the GROUP BY query in SQL. The second method is to define a distance measure for a specific attribute, defined as the clustering key, and cluster by that attribute. Since this is distance based, many standard methods in data mining for clustering can be applied using the second method [125]. The syntax of the cluster operator is given in Table 2.6.

TABLE 2.6: AJAX *cluster* operator syntax

<cluster-operator>	:	create clustering <operation-name> from <predicate-name> [+] <alias-variable> by method <method-name> [with parameters <parameter-name> [{<parameter-name> ... }]] into <predicate-name>

The cluster operator's syntax uses two new constructs, the "by method" clause and the "with parameters" clause. The "by method" clause specifies the method that is to be used to cluster the data while the with parameters clause specifies the parameters for that clause [125].

The Merge Operator

The merge operator defines a method for combining similar tuples into a defining relation. In merging, a single relation defining a set of elements is taken as input and one relation defining a tuple is returned as output. This relates groups and collapses the set of elements within the input relation based on matching criteria [125]. The syntax of the merge operator is given in Table 2.7.

TABLE 2.7: AJAX *merge* operator syntax

<merge-operator>	:	create merge <operation-name> using <predicate-name> [<alias-variable>] [let <let-clause>] [where <where-clause>] <select-into-clause>

The merge operator's syntax uses similar clauses that have been previously explained. However, instead of selecting predicates through the from clause, there is now a using clause. In other operators, the from clause allows the operator's let clause be specified with respect to the predicates of the from clause. In the merge operator, the using clause allows each cluster to be evaluated instead of each predicate [125].

2.3 CASE STUDY IN BIOLOGICAL DATA CLEANING

BIO-AJAX is a data cleaning tool for bioinformatics databases designed to tackle both schema level and data level data quality problems. It uses the conceptual operations developed by Galhardas, et al. [125] presented in the previous section. BIO-AJAX preserves conceptually the operations while modifying the cleaning operations so that they specifically apply to bioinformatics databases needs. These cleaning operations help to reduce a number of the problems mentioned previously that are inherent to bioinformatics databases.

2.3.1 BIO-AJAX Approach

To illustrate BIO-AJAX best, phylogenetic data, specifically phylogenetic trees, were chosen. There are a number of reasons for this choice. First, phylogenetic data tends to be a heterogeneous dataset. The main components are structures in the form of phylogenetic trees modeled in various ways. For a standard keyword database, the problems of data quality and data cleaning have been well explored. However, the problem of cleaning databases containing structural data has not been researched. Moreover, most phylogenetic databases have other data associated with them. Therefore, this leads us to question how does a data cleaning tool operate upon databases with different data? With heterogeneous data come the issues concerning how the tool measures similarity as well as clusters or merges that data. Therefore, phylogenetic data can demonstrate both of these problems and provide an interesting data set for performing cleaning.

To clean biological data, some modifications are needed of the aforementioned framework. While the framework is powerful for many data cleaning problems, there are some extensions that biological data, particular phylogenetic data, need to ensure that the data is fully cleaned. Each of the five operators would need to be extended specifically for the data set. Moreover, there can be multiple extensions of these operators. For example, there are a number of algorithms that can perform matching in any data set including phylogenetic data. Also, for the framework, a sixth operator has been added. This CLASSIFY operator will perform classifications within the database and clean it accordingly.

2.3.2 BIO-AJAX Operators

The BIO-AJAX methodology includes a collection of powerful operators. Here we present the fundamental operators along with important aspects to consider in the biological data cleaning process.

MAP Operator

The mapping operation can help with a number of problems inherent within bioinformatics databases. Mapping is key to biological schema cleaning and biological data cleaning. First, the mapping operation can help with transferring legacy data into new schemas. Second, it can help populate the database with data from other databases by transforming the second database schema into the first database schema. Finally, if there is a schema level problem, mapping can map the data in the dirty schema into a better schema.

Mapping Phylogenetic Trees

One common problem with phylogenetic trees within databases is that the trees can be stored in various formats. Therefore, one possible purpose for mapping a data set of trees is to map them into the same format. If the trees are in different formats, a number of problems can occur. First, to perform any knowledge discovery or even comparison operations on the data set, ideally the data should be in the same format. If not, any tool written to process the data will need to be able to process all formats of the data. However, this method does not manage possible future cases. If the database modifies the format even slightly, any new data will be lost. Mapping the format of the trees helps to accomplish a number of data cleaning objectives. First, it allows for legacy formats within the database to be standardized. This allows all data to be available for any knowledge discovery tool written for that specific format. Second, any data that is not in a required format can then be mapped onto that format. Third, if phylogenetic databases exchange data, mapping can help integrate new data into the pervasive format within the database. Finally, if a database needs to change or update to a new format, the legacy data can be easily updated as well.

Example 1: One common procedure within phylogenetic databases is to extract trees for a knowledge discovery system, such as a comparison tool or a drawing tool. Since data files within these databases tend to be large, it is extremely useful to extract only the data needed for cleaning. Moreover, the trees within these files can be in different formats. The mapping operation can be used to extract the relevant data, such as the tree structure, from the dirty data table for further processing. Based on the grammar in the previous section, the query shown in Table 2.8 would extract a key from the data using an atomic function as well as the tree. The extractPhyloTree function can extract the tree from the data. This operation may require extracting the tree from files with other information. This also may require that the tree be formatted separately from its original format so that it can be mapped according to the new schema.

TABLE 2.8: BIO-AJAX *create mapping* query

CREATE MAPPING ExtractPhyloTree
FROM dirtyPhyloData dph
LET tree = extractPhyloTree(dph.data),
key = generatePhyloKey(dph.data)
WHERE PhyloData.tree < > null
{SELECT dph.key AS phyloKey, dph.tree AS Tree
INTO DirtyPhyloTreeData
CONSTRAINT NOT NULL phyloKey}

VIEW Operator

The view operation should be standard with what has already been discussed since it is a method for querying the database. For example, the view operator can be used to display the tables generated with mapping operator.

Example 2: The query shown in Table 2.9 displays the phylogenetic information extracted by Example 1.

TABLE 2.9: BIO-AJAX *create view* query

CREATE VIEW viewTrees
FROM DirtyPhyloData d1, DirtyPhyloData d2
WHERE d1.phyloKey = d2.phyloKey
{SELECT d1.phyloKey AS key, d1.tree AS tree INTO Trees
CONSTRAINT NOT NULL tree}

MATCH Operator

The matching operator has many features that can help facilitate data cleaning. First and foremost, it is an excellent tool for detecting duplicates or records whose semantic contents are so similar that in essence they are duplicates. For many of the biological fields that have readily accessible databases, exact matching and similarity matching are well-explored research areas. For example, for a database containing nucleotide or amino acid sequences, sequence alignment algorithms such as BLAST, FASTA and CLUSTAL-W can tell a user how similar two sequences or a set of sequences are. Moreover, while these areas are well explored, most do not include similarity measures that can be considered perfect for every comparison.

A large problem within bioinformatics research is the idea of detecting synonymous data. Since biological data can be complex, incorporating structures as well as text, detecting entries that are not identical syntactically but are identical semantically is a large problem. For example, it is common for the more heavily studied species to have multiple identification names. Most species at least have a scientific name and a language-specific vernacular name (e.g. human and *Homo sapiens*, fruit fly and *Drosophila melanogaster*). This problem can be compounded by a number of factors. First, the scientific or Linnaean names for organisms can be inadequate for describing a species. Many species were classified before genetic sequencing and other more quantitative methods were developed to classify them. With the use of the more qualitative methods, separately classified species ultimately were treated as variations within the same species. Since many phylogenetic databases have legacy data and there are many sources for phylogenetic information, standardizing nomenclature or the naming mechanisms used to identify a species throughout a set of data and checking nomenclature within a knowledge base can help clean the data significantly. By standardizing nomenclature, comparisons and other knowledge extraction tasks can be performed much more effectively and efficiently.

Besides nomenclature, matching can also be used for error detection. Biological data presents a number of causes for errors. Like any other database, bioinformatics databases can easily suffer the same problems of improper input, spelling mistakes and non-standard abbreviations. However, if used with a knowledge base, matching can be a very effective method for detecting errors [228].

A knowledge base for species can be preliminarily developed from readily available web databases. For example, NCBI provides a taxonomy tool that may be exploited into a knowledge base. The NCBI taxonomy tool, located at

http://www.ncbi.nlm.nih.gov/Taxonomy/taxonomyhome.html

gives a great amount of information about any given species. By querying these tools with a given species name, the various nomenclatures about the species can be extracted from the resulting Web pages. Once the nomenclature is extracted from the Web pages, it can be used to standardize the nomenclatures in the phylogenetic trees.

Also, duplicate or similarity detection is very important in phylogenetic studies. When a duplicate is detected, one of many different events may have occurred within the database. First, there could be a duplicate record and the record should be removed. Another case may be two identical trees by the same author in two different studies. The curator of the database may want to decide how to handle this duplicate information. Next, there may exist duplicate trees within the database by different authors using different reconstruction algorithms. This could be a potentially important finding since most algorithms that create phylogenetic trees do not perform the exactly same operations as other reconstruction algorithms. This information could also be helpful to the users of the database if they want to see similar trees.

The matching operator can incorporate a number of matching algorithms that are native to biological data sets or specific phylogenetic tree data sets. There are also matching algorithms, while not native to bioinformatics databases, customarily used on keyword databases that can be used to process phylogenetic trees, especially the metadata associated with the trees. However, these algorithms must be used so that a match can be detected on the biological or phylogenetic data. For example, the matching operator can include sequence alignment, structure comparison and a thesaurus to handle the species synonymy problem. With the various terminologies as well as the different methods for analyzing the structures involved within biological data sets, it is very possible to have duplicate records that are not evident through keyword comparison.

Matching and Phylogenetic Data

A number of similarity and distance measures for phylogenetic trees have been developed. Various algorithms for tree matching [383] and tree reconstruction have been studied. Different theories, when applied to finding the phylogenetic relationships of the same set of species, often result in different phylogenetic trees. To determine how much two theories have in common is a fundamental problem in computational biology and in phyloinformatics.

A comprehensive algorithmic and software tool in this field is the COMPONENT package [283]:

```
http://taxonomy.zoology.gla.ac.uk/rod/cpw.html
```

COMPONENT provides several ways of finding the consensus of two phylogenetic trees. The first is to see whether they have similar quartets based on adjacency relationships among all possible subsets of four leaf species. The similarity is then computed as the proportion of quartets shared by the two trees.

Partition distance treats each phylogenetic tree as unrooted and analyzes the partitions of species resulted from removing one edge at a time. By removing an edge in a tree, we can partition that tree. The difference between two trees is defined as the number of edges in a tree for which there is no equivalent (in the sense of creating the same partitions) edge in the other tree.

The maximum agreement subtree between two phylogenetic trees T_1 and T_2 is a substructure on which the two trees are the same. Commonly such a subtree will have fewer leaves than either T_1 or T_2. By contrast, a consensus tree has the same number of leaves as the original trees T_1 and T_2, assuming those trees have the same set of species. Consensus trees should be used gingerly, however, because a consensus tree is not a phylogeny unless the two trees are isomorphic. Instead, consensus trees present a convenient way to summarize the agreement between two or more trees. Consensus trees can be formed from cluster methods (strict, majority rule, semi-strict, or Nelson) or by intersection methods (Adams consensus) [283].

The last dissimilarity measure implemented in COMPONENT is the nearest neighbor interchange (NNI) distance. Given two unrooted unordered trees T_1 and T_2 with the same set of labeled leaves, their NNI distance is the minimum number of NNI operations needed to transform T_1 to T_2. Finding the NNI distance between two trees is NP-hard [132]. Several approximation algorithms to calculate NNI distance are implemented in COMPONENT [55, 283].

The following example demonstrates how matching can be performed using the research from [383] that was eventually implemented into the ATreeGrep algorithm for searching unordered trees [333].

Example 3: In Table 2.10, two trees are compared and the degree of similarity can be controlled through the third parameter of MAXDIST in the WHERE clause.

An in-depth case study of phylogenetic tree matching methods is presented in Chapter 4 of this book.

CLUSTER Operator

The clustering operation can perform similar duplicate detection functionalities that the matching operator can perform. Conceptually,

TABLE 2.10: BIO-AJAX *create matching* query

CREATE MATCHING MatchDirtyPhyloTrees
FROM DirtyPhyloTreeData d1, DirtyPhyloTreeData d2
LET distance = ATreeGrepDiff(d1.tree, d2.tree)
WHERE distance < MAXDIST (d1.name, d2.name, 0)
INTO MatchTrees

the cluster operation organizes a set of elements in a relation by either their value or their distance from one another. Both value and distance can be defined for any given data set, especially biological data, so that this operation can be performed. For example, for protein data, clustering can be performed for proteins that have similar structures or amino acid sequences. For nucleic databases, clustering can occur on sequences that have specific gene or protein or segments of DNA that are similar. For phylogenetic data, this can be performed on a set of phylogenetic trees that have a similar structure.

Clustering and Phylogenetic Data

Clustering phylogenetic data has been an important function for a very long time. By clustering phylogenetic data, relationships between species can be discovered. Also, through clustering phylogenetic trees, comparisons of various reconstruction algorithms can be made. By clustering phylogenetic trees specifically, methods for creating the trees or reconstruction algorithms can be compared. Clustering allows scientists and database managers to see what data can be considered similar and what is possibly an aberration within a database. This can help with error detection since clustering can highlight outliers that can then be analyzed and cleaned appropriately. Moreover, through these comparisons, various identifying aspects of the trees can be learned. For example, by clustering similar trees, scientists can learn how closely species or groups of species are related to each other. Reconstruction algorithms can also be evaluated. Clustering can demonstrate the effectiveness of an algorithm as well as its flaws. For example, for a given set of peer-reviewed trees, clustering can find the common structure between two trees. Moreover, if a user would like to compare his or her tree against similar trees, clustering can help the user see what trees his or her tree should be compared against. If the tree is then compared to the database and the new tree is not included within the cluster, the result could indicate a problem with the reconstruction algorithm or tool that created the tree [348].

Clustering phylogenetic trees is a complicated topic. Currently, the most popular approaches use a combination of common clustering methods and consensus trees. In these phylogenetic tree clustering algorithms, standard clustering algorithms such as k-means and agglomerative clustering can be applied to the phylogenetic trees. Once the trees are in clusters, consensus trees can be formed from these trees to represent the clusters. A consensus tree is a tree T where, given a set S of trees, all edges in T are contained within every member of S. The formation of a consensus tree creates a representative for a clustering of phylogenetic trees. Consensus trees can be used without first clustering through the standard algorithms; however this creates one tree for the entire data set [348].

Example 4: This example shown in Table 2.11 clusters phylogenetic trees. The trees are clustered based on similarity. This can be highly useful for comparing phylogenetic trees and their methods of construction.

TABLE 2.11: BIO-AJAX *create clustering* query

CREATE CLUSTERING clusterTreesbyStockhamMethod
FROM MatchTrees
BY METHOD StockhamMethod
INTO clusterTrees

Outlier Operator as Part of CLUSTER Operator

With the aforementioned operators, it is possible to detect and correct a number of instances where data is considered "dirty." However, errors within the data can still pass unnoticed. Moreover, with clustering and merging operators, it becomes important to analyze the results of the clustering operation in a more in-depth manner. While clustering can help detect similar or duplicate records, outlier records can also indicate errors in the data.

An outlier is a data point that is different from the rest of the data within a database for some given measure. Outliers can be detected easily within a data set that has been clustered, since the outliers will not be contained in any cluster. If a data point does not belong to any cluster, this means that the data point is not similar to any other data points in the data set.

Concerning phylogenetic data, outlier detection can yield a great amount of information. First, concerning data cleaning, outlier detection can indicate a problem with the data. If a given tree T created with a set of taxa S and a reconstruction algorithm R does not behave in an expected manner, such as falling into clusters with other trees based on S and R, there may be an error with the data. This error can be the result of a number of problems. Some of these may be input errors by the creators, improper instantiation of the reconstruction algorithm, improper application of a reconstruction algorithm, and faulty data used to create the tree. The syntax of the *outlier* operator is given in Table 2.12.

TABLE 2.12: BIO-AJAX *outlier mapping* operator syntax

<outlier-mapping-operator>	:	create <outlier-mapping-name>
		using <predicate-name>
		[<alias-variable>]
		let <let-clause>
		where {cluster-size = 1}
		[<where-clause>]
		<select-into-clause>

MERGE Operator

The merge operation can act similarly to matching and clustering. Since merging occurs based on "value" or "distance" with respect to a given attribute, the operations that can be performed are very similar to those in clustering. However, since the data is grouped and collapsed, this method may be useful for creating consensus trees, supertrees or superstrings. Also, if duplicates are detected within the match phase, merge can be used to reduce a set of identical elements into one element.

Merging phylogenetic trees is commonly done in various applications of research. Usually trees are merged through the use of supertrees [319]. Supertrees are pervasive methods throughout a study of phylogenetics for relating phylogenetic subtrees to each other. The purpose of the supertree is to use a set of phylogenetic subtrees, which may or may not contain the same species, and form one tree that preserves as best as possible the evolutionary relationships between the species within all of the trees [319, 332].

Since scientists studying phylogenetics theorize that life originated with one species and all other species are evolved from that one, the supertree models a possible pattern for this evolution. It specifies the relationship between species, and also specifies the relationship between trees. While supertrees can be used to merge any set of trees, they offer an interesting capacity for phylogenetic data cleaning in that they can act as a merge function for phylogenetic data. If two tree records are detected to be similar enough for merging, supertree algorithms can be used to merge the trees within the phylogenetic records [319, 332].

Example 5: This example shown in Table 2.13 merges similar data within a phylogenetic database. Merging phylogenetic data has a number of uses. First, if exact matches are found and represent a duplication error, the duplication can be eliminated. Second, if the database wants to condense similar trees, merging can perform this task. Finally, merging can also facilitate the formation of supertrees.

TABLE 2.13: BIO-AJAX *create merging* query

CREATE MERGING MergeTrees
USING clusterTrees ct
LET tree = getTreeCluster(DirtyPhyloTrees(ct).tree)
key = generateKey()
{SELECT key AS phyloKey, tree AS PhyloTree INTO Tree}

CLASSIFY Operator

The final component for the BIO-AJAX framework is the CLASSIFY operator. In working with biological data, performing classifications is a routine procedure. Classification aids biologists and database curators alike. First, classification offers the ability to discover patterns within the data if no known patterns already exist. Moreover, it also allows us to explore known patterns deeper and possibly discover more patterns. Classification also helps with a data item that has a variety of data quality issues.

Due to the nature of the biological data, especially its enormous size and complexity, pattern recognition within the data set becomes extremely important. Through classifying the data, previous observations can be confirmed and enriched. If a new data item fits the classification model within given parameters, it adds credence to the classification model. On the other hand, if it does not fit the model, it can indicate an error in the data or in the classification model. Adding to the knowledge of the patterns within a given data set helps biologists to better understand the underlying mechanisms within biology and better model the data with respect to these mechanisms.

In addition, classification can help with a number of data quality issues. Consider, for example, a new protein whose function is not fully understood yet. Through classifying the protein, the protein that is most similar to it can be found. If the proteins are very similar, some aspects of the information stored about the known protein may be also connected to this new protein, which helps to improve the consistency of the data. Moreover, it helps to disseminate metadata, which is currently a key concern for all bioinformatics databases.

Table 2.14 shows the syntax for the CLASSIFY operator. The syntax is similar to that of the MATCH operator since their functions are similar. However, classification helps to identify a data item as opposed to comparing it against other data items within the repository to detect similarity.

TABLE 2.14: BIO-AJAX *classify* operator syntax

<classify-operator>	:	create classification <operation-name> from <predicate-name> [+] <alias-variable> [{<predicate-name> [+] <alias-variable>...}] [let <let-clause>] [where <where-clause>] into <predicate-name>

Concerning phylogenetic data, there are many instances where classification may be needed. As to nomenclature problems, classification can offer the ability to standardize a set of nomenclature to a specific format by revealing whether it already exists in a certain format. Also, classification can help in analyzing phylogenetic tree structures. If a reconstruction method is unknown or there is inconsistent data, classification can aid in solving these problems. It can also identify characteristics of reconstructions and any abnormalities within a construction.

Example 6: Table 2.15 shows classifying a phylogenetic tree that is missing reconstruction method implementation.

TABLE 2.15: BIO-AJAX *create classification* query

CREATE CLASSIFICATION
classificationTreeReconstructionParsimony
FROM MatchTrees
BY METHOD ParsimonyClassifier
INTO classParsimonyTrees

CHAPTER 3

Biological Data Integration

CONTENTS

3.1   INTRODUCTION

Many data cleaning problems arise from how data in different schemas interact with each other or map to a common view. These problems affect the retrieval recall and precision of a data repository, its ability to interact with other data repositories and the repository's ability to incorporate or modify the model of the data. Whether the cleaning problem arises from one of these problems or other problems where ultimately, data from one set must be mapped to another, the situation can be viewed as a data integration problem.

Data integration has been heavily studied in the past and resulted in a number of innovative methods for the interaction of data among databases, particularly those accessible through the Web [6]. Its applications can include any function from integrating databases so various knowledge discovery tools can be applied to data distributed to various databases to providing users with a simple interface. However, traditional concepts concerning data integration can limit the integration possibilities for bioinformatics databases.

With bioinformatics databases, the integration problem becomes more difficult than creating a seamless view of a set of data for a user. Integration must be considered as a problem affecting interacting databases and also as a problem that affects multiple areas of the database. These areas include submitting new or modified data to the database, having data within a large repository interact with each other and having separate databases present a unified view of data from multiple database projects to a user. Ideally, most bioinformatics databases need to address data integration on at least three levels. These levels include:

- Integrating submission data

- Integrating data currently within the database

- Integrating data with external data repository resources

Integrating data based on these levels offers bioinformatics databases the ability to house data more effectively while giving users more control when researching. It allows scientists to gather their results in one location and standardize the format of their results. It gives curators of these databases a tool that frees them from creating a single centralized database for their data. Finally, it allows independently developed databases to interact with each other, permitting users to take advantage of the abilities of using knowledge discovery tools from multiple databases in one interface.

3.2 GENERAL DATA INTEGRATION

Data integration is a well-known problem within the fields of database management and bioinformatics. Commonly, the data integration problem is seen as the need for combining data residing in different sources into one seamless view for the user [225]. This frees the user to be concerned only with the query rather than locating sources for the data of interest or collating data from multiple sources to get a complete view.

A generic data integration system typically consists of three parts: the global schema, the source schema and the mapping between the global and source schemas [225]. The global schema refers to the combination of the data residing in or originating from various sources. The source schema handles organization of the data obtained from the source data sets. The mapping describes how the data in the source schema participates in the global schema. For the integration system to be acceptable, the integration must satisfy the constraints imposed upon it by the global schema while also satisfying the mapping between the global schema and the source data sets.

Data Integration Methods

The two most common data integration approaches are the "local as view" (LAV) and "global as view" (GAV) methods. Both approaches define methods for creating the mapping between the global schema and the source data sets. In the LAV approach, the global schema is specified independently of the source data sets. The source data sets are seen as views over the global schema. The GAV approach specifies the global schema in terms of the source data sets. Each element of the global schema is seen as a view over data within the source data sets [225].

Both LAV and GAV have disadvantages. In the LAV approach, query processing is difficult. Since the only information known about the data in the global schema is through the views of the sources, which are limited, the data can be incomplete. While GAV does not have this problem, adding new sources or new data to the global schema can become difficult. To help maximize the benefits of the integration, some databases use a combination of these approaches, referred to as GLAV, to create the integration [63, 225]. Researchers in the data integration communities have looked mainly at continuing efforts concerning schema mediation languages to query answering algorithms, query optimization, query execution and create applications using data integration techniques [152, 153, 154]. See [64] for an excellent treatment to this field.

Data Integration Systems

One important data integration system is BibFinder [274, 275]. BibFinder is a citation index for computer science articles, which integrates citation searches from the collection of Computer Science Bibliographies (CSB), the DBLP Computer Science Bibliography, Network Bibliography, ACM Digital Library, ACM Guide, ScienceDirect, IEEE Xplore, CiteSeer, and Google. BibFinder is a testbed for the research of the Havasu Project [274] and is innovative in that it addresses many complex issues of data integration. The Havasu Project develops automated methods for executing a query over multiple, heterogeneous databases simultaneously, accounting for the cost and optimization issues for the query. The research from the Havasu Project has also been extended into BioHavasu, a mediator for biomedical genomic data. However, this mediator currently requires that the user knows the gene accession number to use the system.

Another important integration project is a component of the Foundations of Data Warehouse Quality Espirit Project [184]. This project involves a number of research centers and universities in Europe laying a foundation for developing advanced database facilities, such as deeper models, more complex and interesting indices as well as semantic services for data warehouses [184]. Data integration techniques have been used to help further improve data quality, mainly through integrating original sources into the warehouses.

The database group at the University of Washington works on a number of projects regarding integration. For example, the Piazza project investigates integrating peer-to-peer resources [154] using semantic mappings. It employs both LAV and GAV methods, using XML Piazza as an extension of the Sagres system, which looks at data sharing among invisible and ubiquitous computing devices [183]. Another significant project is the Tukwila data integration system, which offers state-of-the-art network query processing. This group also works on BioMediator, which helps integrate genomic resources.

Besides the research at the University of Washington, a number of groups are looking at XML and Semantic Web technologies for data integration. In order to offer all of the functionalities the Semantic Web proposes, there is a need for integration. Integration technologies for the Semantic Web include XML and the various communication layers associated with the Semantic Web language tower such as RDF and DAML+OIL. Doan, et al. [6] present an excellent review of data integration systems developed by or used in industry and academia.

3.3   TOPICS IN BIOLOGICAL DATA INTEGRATION

Biological data integration problems have been discussed in many papers [93, 213]. Some of the initial successful biological data integration projects dealt with biomedical periodical data. A couple of successful projects based on this type of integration are BioKleisli [94], GeneScene [228, 229] and MedTextus [228, 420]. In [6, 212], the authors discuss XML as an excellent medium for interchanging biological data. From the base of XML and Semantic Web technologies, numerous integration projects were designed. Within bioinformatics, some noteworthy projects include Gene Ontology [43], TAMBIS project [347], BioFAST and Biological Integration System [211].

Within bioinformatics databases, most freely available World Wide Web repositories have some level of integration. Archival databases tend to have multiple levels of integration. If an archival database is partnered with mirror sites, the sites are usually synchronized with each other regularly. Other databases that are not mirrors also connect in a less rigid manner, using Internet connections to query each other's databases. For example, NCBI [7, 37, 350, 360] uses "link out" to other prominent repositories so that the user can obtain more information on a particular subject.

Another interesting movement within this field is to use integration to unify the repositories. The Protein Data Bank (PDB) [38, 390, 391] uses various technologies to integrate new data into its repository as well as integrate legacy data and new data into one unified view of the repository. The Protein Information Resource (PIR) [404] also employs integration technologies to help search its various repositories. iProClass gives the user a unified search interface for the data found in the PIR-PSD, Swiss-Prot, and TrEMBL databases [175]. Most of the integrated bioinformatics database systems address the concerns of providing the user with as much data as possible. However, integration can facilitate other areas of importance to users. Bioinformatics databases have become more than data repositories and serve as analysis tools. Therefore, issues concerning what data a bioinformatics database can recall and what tools can be applied to that data are of great importance.

Bioinformatics databases, like most databases, are subject to many challenges. However, due to the nature of the biological data, usually these challenges become complex, thereby excluding the more traditional methods for solving the problems, complicating the integration problem. The data integration problem within most databases concerns combining data from various sources to represent a unified view to the user of the data within those databases. Applying this definition to biological data, there are three specific areas to which data integration can then be applied. These areas are integrating scientific data, integrating databases within one repository and integrating multiple repositories [162].

3.3.1 Integrating Scientific Data

The first area, integrating the data scientists submit to bioinformatics databases, is always an interesting problem. Issues arise from wanting to give scientists the freedom and the flexibility to specify their results descriptively while also controlling the submission so that integrating the data into the database is possible. If biologists are too limited during the submission process, key elements of their results may be expressed within the database improperly. Also, since knowledge within biology is dynamic, with new understanding of the data developing constantly, any information not included can affect results in the future. However, a number of problems can develop by not controlling the submissions into the database. First, by not controlling vocabulary, there will be very little consistency between entries. This can cause many problems, especially concerning retrieval of the data submitted as well as applying knowledge discovery tools to the data.

If data are inconsistent, it becomes harder to retrieve since two authors may use a different set of phrases to describe similar concepts. For example, two researchers may enter nucleotide information about the species *Dario rerio*. One enters the data as *Danio rerio* while the other enters the data *D. rerio x1*, where *x1* indicates the specific organism the sequence was obtained from. Without some consistency mechanism, whether it is a controlled vocabulary that restricts the user to a specific

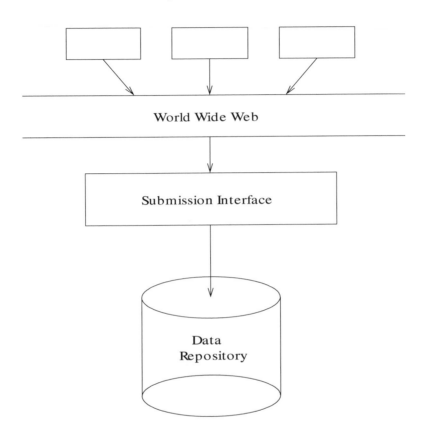

FIGURE 3.1: Scientific data integration.

nomenclature format or a thesaurus that reconciles the input, the later study would not be returned for anyone querying *Danio rerio* nucleotide sequences. Moreover, without any control mechanisms placed upon the submissions, it becomes very difficult to check for errors. Submissions presented within controlled vocabularies can greatly affect the precision of the database. The synonymy and polysemy problems become difficult to control, and query mechanisms become very complex.

Figure 3.1 demonstrates this type of integration. A researcher would submit data to the data repository. This data would most likely be submitted through the World Wide Web via some submission interface. Once the data is formatted through the interface, it is stored within the database.

Currently, most bioinformatics databases use Web-based methods for accepting data submissions. These methods include a Web page interface that guides researchers through the submission process and email. The interface is generally a Web form through which a researcher can submit data. If the researcher does not feel comfortable with this method, many bioinformatics databases also offer the ability to email the data to them. If the data is complex, sometimes a curator will review the submission manually and then insert it into the database. If the Web form is not complex or has a strictly controlled vocabulary, it may be inserted automatically.

3.3.2 Integrating Databases within One Repository

The second form of integration concerns integrating data within one repository. For some of the larger bioinformatics databases, multiple management systems and data schemas are used to store the information within the scope of the databases. This type of database architecture is used for a number of reasons. Some databases house large amounts of legacy data. When there is a schema redesign, the legacy data does not always conform to the new database schema. Therefore, the legacy schema can retain the legacy data and a new schema can be developed for the new data. An integrated schema can then be imposed upon this data. Also, since biological data is a complex data set, some database management systems (DBMSs) offer better functionalities for storing specific data than other systems.

Figure 3.2 demonstrates integrating within one repository. It shows a collection of data sets housed within DBMSs. These databases then interact through some integration technology. Possible integration technologies include relational database tools similar to the one used in PIR's iProClass [175]. Other possibilities include integrating the databases using Web technologies, as if each database occupied its own separate repository. The user would then have a unified view of these databases.

3.3.3 Integrating Multiple Repositories

The third form of integration concerns integrating data from multiple bioinformatics databases to allow a user to search for any topic. The interface would then return data stored in databases participating within the tool in a unified manner. There are a number of projects that partially implement this type of integration [94, 349]. For example, the EpoDB project [349] uses similar techniques. The EpoDB is a database designed for analyzing the genes expressed during vertebrate erythropoiesis. This database creates a conceptual database using Sybase, ex-

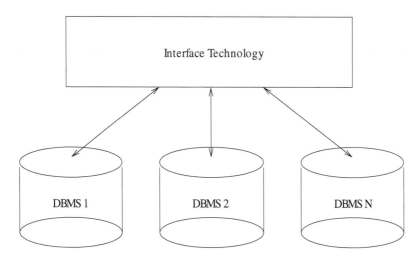

FIGURE 3.2: Data integration in a single repository.

tracting data from multiple sources including GenBank and Swiss-Prot [349].

Figure 3.3 demonstrates this type of integration. It assumes that the user is interacting with the integration view of the database through a Web interface. However, this framework could be modified to represent any network medium. Both the user and the databases that take part in the integration interact with the integration interface. The user connects to the interface through the Web, represented by the dark line. The data repositories also connect to this integration interface that is responsible for query processing. This connection is represented by the dashed line. The data repositories then interact with the integration interface to create a unified view for the user.

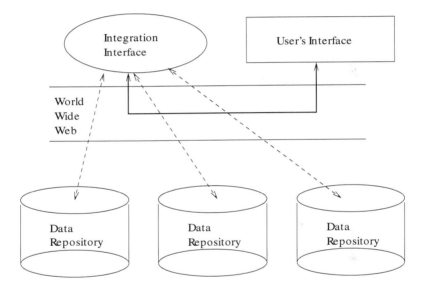

FIGURE 3.3: Data integration in multiple repositories.

CHAPTER 4

Biological Data Searching

CONTENTS

4.1 INTRODUCTION

As presented in Chapters 1 through 3, digitized biological data are plentiful. These data are categorized by life species, gene, RNA function and evolutionary history. This vast resource of biological data is a treasure trove ready to be tapped. Knowing where and how to find useful biological information is important.

In this chapter, useful utilities for conducting biological searches are discussed. BLAST [13] is usually among the first biological tools learned and used by researchers for basic nucleotide or protein sequence searching. The UCSC Genome Browser [199] gives a broader perspective of differences between similar or dissimilar organisms. At even higher level [380], searching for the evolutionary history of a creature can reveal unexpected similarities to other creatures. Finally, RKalign [340, 341] is a valuable tool when searching for an RNA motif, a pseudoknot, which is known to play a key role in numerous critical cellular functions.

4.2 BIOLOGICAL DATA SEARCHING USING BLAST

The Basic Local Alignment Search Tool (BLAST) finds regions of local similarity between sequences [13]. BLAST, first used around 1990, is one of the best and widely used bioinformatics tools available. The program compares nucleotide or protein sequences to sequence databases and calculates the statistical significance of matches. BLAST can be used to infer functional and evolutionary relationships between sequences and help identify members of gene families. With each BLAST search result, the statistical significance of the result is provided with a score.

4.3 BIOLOGICAL DATA SEARCHING USING THE UCSC GENOME BROWSER

The UCSC Genome Browser [199] provides search results in the context of the complete genome of an organism. Originally developed for the human genome [217], the browser now houses genomes from numerous organisms and is available as an interactive Web interface site. Genome sequence data from a variety of vertebrate and invertebrate species and major model organisms are available for biological data searching [44, 86, 140, 150, 298, 346, 362].

As vertebrate genome sequences near completion and research refocuses to their analysis, the issue of effective genome annotation display becomes critical. A mature Web tool is used for rapid and reliable display of any requested portion of the genome at any scale, together with several dozen aligned annotation tracks (*http://genome.ucsc.edu*). This genome browser displays assembly contigs and gaps, mRNA and expressed sequence tag alignments, multiple gene predictions, cross-species homologies, single nucleotide polymorphisms, sequence-tagged sites, radiation hybrid data, transposon repeats, and more as a stack of coregistered tracks. Text and sequence-based searches provide quick and precise access to any region of specific interest. Secondary links from individual features lead to sequence details and supplementary off-site databases.

4.4 CASE STUDY IN PHYLOGENETIC TREE DATABASE SEARCH

4.4.1 Background

This case study presents a tool and similarity measures for searching in phylogenetic databases [329, 380]. Scientists model phylogenetic relations using unordered labeled trees and develop methods for constructing these trees [39, 65, 117, 148, 193, 385]. Different theories concerning the phylogenetic relationship of the same set of species often result in different phylogenetic trees. Even the same phylogenetic theory may yield different trees for different orthologous genes. With the unprecedented number of phylogenetic trees based on these various theories, the need to analyze the trees and manage phylogenetic databases is urgent [296].

One important need is to be able to compare the trees, thus possibly determining how much different theories may have in common [59, 83, 90, 194, 365]. The common portion of two trees may represent added support for the phylogenetic relationship of the corresponding species.

One motivation for studying the tree matching problem comes from the development of tools for analyzing the phylogenetic data. One tool, presented here, is a system for searching a database of phylogenetic trees. Given a query or pattern tree P and a set of data trees $\mathcal{D}$, the tool is able to find near neighbors of P in $\mathcal{D}$ where the similarity scores between near neighbor trees and P are greater than or equal to a user-specified cutoff value σ. Central to the tool's search engine is an algorithm for computing the similarity score from P to each data tree D in $\mathcal{D}$. Data consists of the phylogenetic trees stored within the widely used phylogenetic information database TreeBASE [296, 318] (http://www.treebase.org). These phylogenetic trees model the evolutionary history among life forms. The sampled life forms, whose biological characteristics are used to infer their evolutionary history, usually appear as leaf nodes. Each internal tree node represents an inferred ancestor of the organisms represented by its child nodes. There can be multiple levels of ancestors, with multiple organisms sharing the same ancestors. A similarity measure [379] compares phylogenetic trees that satisfy the following properties:

- Each tree is rooted and unordered, i.e., the order among siblings is unimportant, and no weight is assigned to any edge of the tree.

- Each leaf node has a label and that label appears only once in the tree, although it may appear in other trees.

- Each non-leaf node either has a label that appears nowhere else in the tree or has no label; an unlabeled internal node stands for an unnamed evolutionary unit.

- Each unlabeled internal node has at least two children.

These properties characterize many trees in TreeBASE and those generated by modern tree reconstruction programs. This case study extends the similarity measure [379] to compare unrooted phylogenetic trees as well as weighted trees, i.e., trees whose edges have weights. The similarity measure and search algorithms for rooted trees [379] are first reviewed. Next, extensions for unrooted trees and weighted trees are presented. Then, a comparison of the similarity measures presented in this case study against existing tree metrics is shown. Finally, performance results for near neighbor searching are detailed.

4.4.2 Methods

Up and down operations

Unless otherwise stated, discussed here are rooted, unordered trees satisfying the four properties described in Section 4.4.1. Two types of operations, up and down, between any two nodes in a tree are defined. These operations are intended to capture the hierarchical structure in the tree. If v is a child node of u, $v \uparrow u$ represents an up operation from v to u, and we use $u \downarrow v$ to represent a down operation from u to v. Then, for any pair of nodes m, n in the tree T, we can count the number of up and down operations to move, say a token, from m to n.

To illustrate the calculation of up operations or UpDistance, Figure 4.1 shows a seven-node tree with node Z at the root level and node A at the leaf level. Since there are two levels between leaf node A and root node Z in this tree, the number of up operations or the UpDistance from node A to node Z is 2. That is, it takes two up operations ($A \uparrow Y$ and $Y \uparrow Z$) to go from A to Z in the tree.

As another example, consider the tree in Figure 4.2 and the fox and rabbit nodes in the tree. It takes two up operations (fox $\uparrow$ carnivore and carnivore $\uparrow$ mammal) and one down operation (mammal $\downarrow$ rabbit) to go from fox to rabbit in the tree. As another example, it takes one up operation (dog $\uparrow$ carnivore) and one down operation (carnivore $\downarrow$ fox) to go from dog to fox in the tree.

Updown matrix

Given a tree T, two matrices can be built and referred to as the up matrix U and the down matrix D. They consist of integer values where $U[u, v]$ represents the number of up operations from node u to node v and $D[u, v]$ represents the number of down operations from u to v in the shortest path connecting u and v in T. Note that $U[u, u] = D[u, u] = 0$

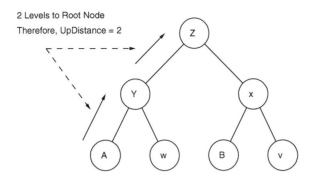

FIGURE 4.1: UpDistance: Given this arbitrary seven-node tree figure, the number of up operations or UpDistance from leaf node A to root node Z is 2. This is illustrated by two arrows near the tree edges (A, Y) and (Y, Z).

for any node u in T. Figure 4.3 shows a tree T and its up and down matrices, U and D, respectively. Notice that one of the internal nodes of T in Figure 4.3(i), i.e., the parent of b and c, does not have a label. An unlabeled node does not appear in the matrices. It can be seen that from matrix U we can obtain matrix D, and vice versa. Matrix D is the transpose of matrix U, and vice versa. Therefore, only matrix U is used throughout this section and is referred to as the Updown matrix. The Updown matrix of a tree T describes the structure of tree T. Computing the Updown matrix for a tree T requires $O(N^2)$ time where N is the number of nodes in tree T.

Updown distance

In general, when using a search engine, if the user inputs a query tree with three nodes fox, dog and tiger plus their parent node mammal, the user often expects to see data trees in search results containing these nodes. If the user doesn't want to see a search result containing, for example, a node, tiger, he or she can simply input a query tree having fox, dog and mammal only. This implies that in designing a search engine

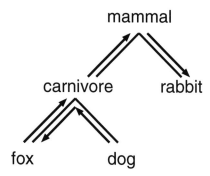

FIGURE 4.2: Up operations and down operations: Arrows show that it takes 2 up operations and one down operation to go from the leaf node labeled fox to the leaf node labeled rabbit. Arrows also show that it takes one up operation and one down operation to go from the leaf node labeled dog to the leaf node labeled fox.

and a similarity or distance measure, the following two criteria should be considered together:

- Whether all, or at least most of, the labeled nodes of the query tree P occur in a data tree D

- To what extent the query tree P is similar or dissimilar to the data tree D in structure

With these criteria in mind, nodes in tree D that match nodes in tree P are sought when comparing tree P with tree D. Specifically, let V_P be the set of labeled nodes in tree P and let V_D be the set of labeled nodes in tree D. Let U_P represent the Updown matrix of tree P and let U_D represent the Updown matrix of tree D. Let I denote the intersection of V_P and V_D; let J denote the difference $V_P - V_D$.

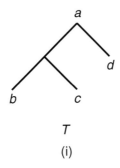

$$T$$
$$(i)$$

$$
\begin{array}{c@{\quad}cccc}
 & a & b & c & d \\
a & 0 & 0 & 0 & 0 \\
b & 2 & 0 & 1 & 2 \\
c & 2 & 1 & 0 & 2 \\
d & 1 & 1 & 1 & 0
\end{array}
$$

$$U$$
$$(ii)$$

$$
\begin{array}{c@{\quad}cccc}
 & a & b & c & d \\
a & 0 & 2 & 2 & 1 \\
b & 0 & 0 & 1 & 1 \\
c & 0 & 1 & 0 & 1 \\
d & 0 & 2 & 2 & 0
\end{array}
$$

$$D$$
$$(iii)$$

FIGURE 4.3: (i) an arbitrary 5 node tree, T, with 4 of the nodes labeled a through d; (ii) Up matrix U representing the number of up operations between any two labeled nodes in T; (iii) Down matrix D representing the number of down operations between any two labeled nodes in T.

The *Updowndistance* from tree P to tree D, denoted $Updowndist(P, D)$, is defined in Equation (4.1).

$$Updown\_dist(P,D) = \sum_{u \in I} \sum_{v \in I} |U_P[u, v] - U_D[u, v]| + \sum_{u \in J} \sum_{v \in J} U_P[u, v]$$

$$(4.1)$$

The similarity score from tree P to tree D, denoted as $USim(P, D)$, is calculated by Equation (4.2).

$$USim(P,D) = \left(1 - \frac{Updown\_dist(P,D)}{\sum_{u \in V_P} \sum_{v \in V_P} U_P[u, v]} \right) \times 100\% \qquad (4.2)$$

The similarity score from tree P to tree D is a measure of the topological relationships in tree P that are found to be the same or similar in tree D. If tree P and tree D are the same or we find a substructure in tree D that exactly matches tree P, then $USim(P, D) = 100\%$. On the other hand, if tree P and tree D do not have any labeled node in common, $USim(P, D) = 0$. The time complexity of the algorithm for computing $USim(P, D)$ is $O(M^2 + N)$ where M is the number of nodes in tree P, and N is the number of nodes in tree D.

Least common ancestor

For any tree satisfying the four properties described in Section 4.4.1, any two distinct nodes of the tree share a least common ancestor (LCA). In evolutionary terms, LCA refers to the most recent species from which two different species evolved. The LCA concept is shown in Figures 4.4 and 4.5. Note in Figure 4.4 the point in the tree identified as $LCA(a, e)$. This shows the least common ancestor of nodes a and e. Similarly, note in Figure 4.5 the point in the tree identified as $LCA(korrae, agrostis)$, also the root of the tree. This shows the node representing the least common ancestor of nodes $korrae$ and $agrostis$.

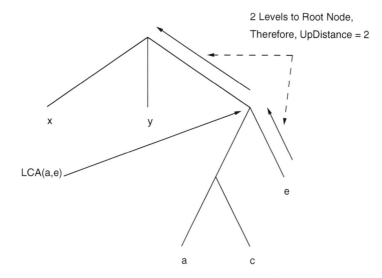

FIGURE 4.4: An arbitrary tree with eight nodes having five nodes labeled a, c, e, x and y, and three unlabeled nodes, including the root node. The UpDistance from the node labeled e to the root node is 2 since there are two up operations needed to go from node e to the root node. The least common ancestor (LCA) of nodes a and c is shown and is one of the three unlabeled nodes.

Tree reduction

Figure 4.6 shows a query tree P and a data tree D that satisfy the four properties described in Section 4.4.1. In the biological sense, when comparing tree P with tree D, the similarity score $USim(P, D)$ should be 100%. Motivated by this example, a data tree reduction technique is incorporated into the structural searching algorithm, which works as follows.

Consider a query tree P and a data tree D and their Updown matrices. Find the column and row indices of the nodes in the intersection of V_P and V_D. Mark those matching nodes in tree D with asterisks. If two distinct nodes of tree D are marked, their least common ancestor (LCA) is also marked. Then consider the reduced data tree D' of tree D that

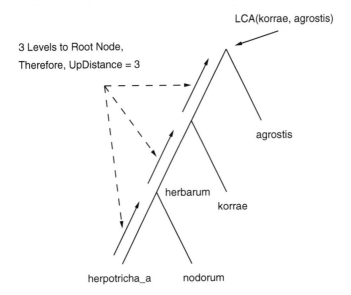

FIGURE 4.5: An arbitrary tree with seven nodes having five nodes labeled agrostis, herbarum, herpotricha_a, korrae and nodorum, and two unlabeled nodes, including the root node. The UpDistance from the node labeled herptotricha_a to the root node is 3 since there are 3 up operations needed to go from node herpotricha_a to the root node. The least common ancestor (LCA) of nodes korrae and agrostis is shown to be the root node and it is one of the two unlabeled nodes.

contains only the marked nodes. Equivalently, unmarked nodes having only one neighbor are removed to preserve connectedness. The node removal may yield additional unmarked nodes with one neighbor, which are also removed. If an unmarked node n is connected to two other nodes $m1$ and $m2$, n is removed and $m1$ and $m2$ are linked to preserve connectedness. These two operations are repeated until neither operation can be performed. The node removal operation is similar to the "degree-2 delete" operation [382] in which a node can be deleted when its degree is less than or equal to 2. After reduction, the Updown matrices will change and the new matrices are used to calculate the Updown distance and similarity score from tree P to tree D.

In Figure 4.7, (i) shows a query tree, (ii) shows a data tree in which

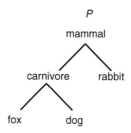

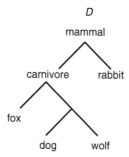

FIGURE 4.6: An example of a query tree P and a data tree D that satisfy the four properties of phylogenetic trees described at the beginning of the case study. In this example, the similarity score $USim(P, D)$ should be 100%.

some nodes are marked, and (iii) shows the reduced tree of the data tree in (ii). In performing a structural search, the algorithm first applies the tree reduction technique to data tree D, and then calculates the similarity score from the given query tree P to the reduced tree of tree D using Equation (4.2). The resulting value is then presented as the similarity score from tree P to tree D.

In Figure 4.7, the similarity score from the query tree in (i) to the reduced data tree in (iii) is 68.42%. Hence, the algorithm displays the data tree in (ii) and indicates that the similarity score from the query tree to the data tree is also 68.42%. This matching technique yields a similar effect as tree matching with variable length "don't cares" [333],

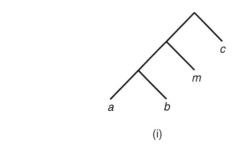

(i)

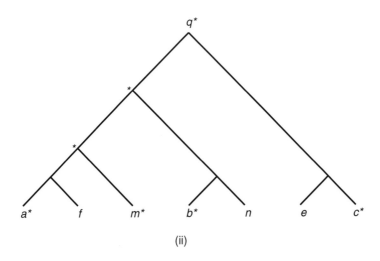

(ii)

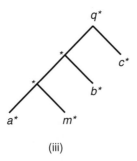

(iii)

FIGURE 4.7: An example of a query tree (i) and a data tree (ii) that both satisfy the four properties of phylogenetic trees described at the beginning of the case study. In the process of tree reduction, nodes on the data tree (ii) are marked after comparison with the query tree (i). The reduced data tree (iii) is shown.

although the approach presented here does not require the user to explicitly specify the "don't cares" in the query tree.

A filter

Given a query or pattern tree P and a database of phylogenies $\mathcal{D}$, it is desirable to find near neighbors of P in $\mathcal{D}$ where the similarity scores between the near neighbors and P are greater than or equal to a user-specified threshold σ. A filter developed to speed the search works as follows. For the database of trees, $\mathcal{D}$, a hash table keyed by a pair of node labels is created with each hash bin containing tree identification numbers. The pair of node labels can be in alphabetical order because $U[u, v] = D[v, u]$ for any pair of node labels (u, v). Given the query tree P, each pair of node labels in P is considered to see which trees of the database the pair belongs to. (This requires time independent of the size of the database.) Data trees are sorted by the number of hits.

When evaluating a data tree D, a lower bound on the Updown distance from P to D is determined by looking at $U_P[u, v]$ where U_P is the Updown matrix of P and (u, v) is a pair in P that is missing from D. The lower bound, denoted Low, is computed by summing $U_P[u, v]$ for all pairs of (u, v) of P missing from tree D. From the lower bound, an upper bound is computed and denoted Upp on the similarity score from P to D, where

$$Upp = \left(1 - \frac{Low}{\sum_{u \in V_P} \sum_{v \in V_P} U_P[u, v]}\right) \times 100\% \qquad (4.3)$$

and V_P is the set of labeled nodes in tree P.

If the upper bound is already smaller than the user-specified value σ, tree D can be eliminated from consideration without calculating the similarity score from tree P to tree D. Furthermore, if a data tree D has a set S of k hits (identified above) and tree D doesn't qualify to be a solution after calculating the similarity score from tree P to tree D, any data tree D' that only has S' of k' hits, where $k' < k$ and S' is a subset of S, will not be a solution and hence can be eliminated from consideration. Experimental results show that this filtering technique works well in practice.

Extensions to weighted and unrooted trees

Some tree reconstruction methods provide information to build a weighted tree where the weight on an edge represents the estimated evolutionary distance between the two nodes connected by the edge [285]. In extending the tree matching algorithm for weighted trees, each up and down operation is associated with a numeric value that represents the weight of the corresponding edge. Instead of having $U[u,v]$ represent the number of up operations from node u to node v, $U[u,v]$ represents the sum of weights associated with the up operations from u to v. Likewise, $D[u,v]$ represents the sum of weights associated with the down operations from u to v. The similarity score between two weighted trees is calculated in the same manner as in Equation (4.2).

Phylogenetic tree reconstruction methods may produce unrooted unordered trees, also called free trees. An unrooted phylogenetic tree specifies only kinship relationships among taxa without specifying ancestry relationships. The common ancestor of all taxa is unknown in an unrooted phylogenetic tree. Each edge in an unrooted tree can be weighted or unweighted.

Assume that T is an unrooted unordered tree. The Additive matrix A for tree T is defined where each entry $A[u,v]$ is the sum of the edge weights on the shortest path connecting nodes u and v in tree T. If tree T is not weighted, $A[u,v]$ is simply the number of edges on the shortest path connecting u and v in tree T (similar to the additive distance for an unrooted tree [39, 60, 384]). Note that when a rooted tree is treated as unrooted, $U[u,v] + U[v,u] = A[u,v]$ for all pairs of (u,v) in the tree, where U and A are the Updown and Additive matrices respectively. Therefore, matrix A can be obtained from U, but the converse is not true. As matrix A is an additive matrix, the four-point condition [60, 415] applies. Hence, an Updown matrix corresponds to a unique Additive matrix which corresponds to a unique tree. This holds for both weighted and unweighted trees.

Now let A_P represent the Additive matrix of the query tree P and let A_D represent the Additive matrix of a data tree D. Let V_P be the set of labelled nodes in tree P and let V_D be the set of labelled nodes in tree D. Let I be the intersection of V_P and V_D; let J denote the difference $V_P - V_D$. The Additive distance from tree P to tree D, denoted as $Add\_dist(P,D)$, is defined as follows [396]:

$$Add\_dist(P,D) = \sum_{u \in I} \sum_{v \in I} |A_P[u,v] - A_D[u,v]| + \sum_{u \in J} \sum_{v \in J} A_P[u,v] \quad (4.4)$$

The similarity score from P to D, denoted $ASim(P, D)$, is calculated as follows:

$$ASim(P,D) = \left(1 - \frac{Add\_dist(P,D)}{\sum_{u \in V_P} \sum_{v \in V_P} A_P[u, v]} \right) \times 100\% \qquad (4.5)$$

The time complexity of the algorithm for computing $ASim(P, D)$ is $O(M^2 + N)$ where M is the number of nodes in tree P, and N is the number of nodes in tree D. It can be shown that for two unrooted trees P and D, whether weighted or unweighted, P and D are identical if and only if the similarity score from tree P to tree D is 100%. This property holds for rooted trees as well.

4.4.3   Results

Comparison of (dis)similarity measures

To evaluate the quality of the similarity measures presented here, the metric $USim$ defined in Equation (4.2) is compared with four widely used tree metrics implemented in the COMPONENT tool (*http://taxonomy.zoology.gla.ac.uk/rod/cpw*). These tree metrics include partition metric (PAR), nearest neighbor interchange metric (NNI), quartet metric (QUA) and maximum agreement subtree metric (MAST). The distribution of the metric values on 945 unweighted rooted trees generated by the COMPONENT tool was evaluated. The query tree was generated randomly; the 945 data trees covered the entire tree space of unweighted rooted trees with 6 labels. The query tree was compared with each data tree to obtain a metric or (dis)similarity value. For PAR, the metric value equals the number of edges in the query tree for which there is no equivalent (in the sense of creating the same partitions) edge in the data tree. For NNI, the metric value equals the number of nearest neighbor interchange operations needed to transform the query tree to the data tree. For QUA, the metric value equals the proportion of quartets shared in the query tree and the data tree. For MAST, the metric value equals the number of leaves removed to obtain a maximum agreement subtree of the query tree and the data tree.

Table 4.1 shows an in-depth comparison between the four widely

TABLE 4.1: Comparison of five studied tree metrics: PAR, MAST, NNI, QUA and WSSP

Metric	Weighted trees?	Internal labels?	Unresolved trees?	Different taxa?	Polynomial computable?	Ref.
PAR	No	No	Yes	No	Yes	Page 2005a
MAST	No	Yes	No	Yes	Yes	Steel and Warnow 1993
NNI	No	No	No	No	No	DasGupta, et al 1995
QUA	No	No	Yes	No	Yes	Bryant, et al 2000
WSSP	Yes	Yes	Yes	Yes	Yes	

used tree metrics and the similarity measures USim and ASim, collectively referred to as WSSP (i.e., Wang-Shan-Shasha-Piel, the designers of the method [380]). In Table 4.1, a Yes value in the "Polynomial computable" column means that there is a polynomial time algorithm for computing the corresponding tree metric and a No value means that computing the corresponding tree metric has been shown to be NP-hard [132]. From Table 4.1 it can be seen that the running time of WSSP is better than NNI (nearest neighbor interchange metric). WSSP can be applied to weighted trees and unweighted trees where trees can be fully resolved or unresolved. It can be used to compare two trees whose internal nodes have labels and whose leaves have different taxa as shown in Table 4.1. In summary, WSSP could be a useful metric in addition to the other excellent ones available.

Efficiency of filter and search method

The WSSP filter technique was tested on synthetic data as follows. One thousand unweighted rooted trees were randomly generated, each tree having 100 nodes. The string labels of nodes were randomly chosen from a dictionary of size 500. The user-specified threshold value σ was set to 60%. In each run, a tree was selected and modified into the query tree and the remaining trees were used as data trees. One thousand runs were performed. Results show that the WSSP filter speeds searches considerably [380]. It was also observed that the running time drops as the user-specified threshold value σ increases. This result was expected since fewer data trees survive the filter when σ becomes larger. Results also show that the WSSP search method scales up well [380]. Running time increases linearly with increasing numbers of trees. Results are consistent with search results using actual phylogenetic trees from TreeBASE.

4.4.4 Conclusion

Unlike many existing metrics [54, 55, 83, 95, 161, 195, 210, 216, 283, 285] designed for comparing two trees possibly with some constraints (e.g., the two trees must have the same set of leaves), the similarity scores described in this case study are mainly developed for near neighbor searching in phylogenetic databases. The similarity scores are not symmetric, i.e., $USim(X, Y) \neq USim(Y, X)$, $ASim(X, Y) \neq ASim(Y, X)$, for any two trees X and Y. The non-symmetry property is good in query-driven phylogenetic information retrieval. This property distinguishes between the situation in which X is a query tree and Y is a data tree and the situation in which Y is a query tree and X is a data tree.

It should be noted that when a substructure in a data tree D exactly matches a query tree P, $USim(P, D) = 100\%$, but the converse is not true. For example, if tree $P = ((a, b), (c, d))$ and tree $D = ((a, b), c)$, the similarity score will be smaller than 100% despite the fact that a substructure of tree P exactly matches tree D. On the other hand, if tree $D = ((a, b), (c, d))$ and tree $P = ((a, b), c)$, the similarity score is 100%. Moreover, the similarity score from tree P to tree D strongly depends on the size of the subset of taxa in the query tree but not in the data tree. The larger the size of the subset of the taxa, the smaller the similarity.

In summary, an approach to near neighbor searching for phylogenetic trees is presented in this case study. Given a query or pattern tree P and a database $\mathcal{D}$ of trees D, the approach finds data trees D in $\mathcal{D}$ where the similarity score of tree P to tree D is greater than or equal to a user-specified threshold value σ. Similarity measures are available for comparing rooted and unrooted trees that are weighted or unweighted. The algorithms presented in this case study have been used for analyzing the structures of phylogenetic trees and for performing structure-based searches in TreeBASE.

4.5 CASE STUDY IN RNA PSEUDOKNOT DATABASE SEARCH

4.5.1 Background

An RNA pseudoknot is formed when a base on a single-stranded hairpin or internal loop is paired with a base outside that loop [104, 297, 405]. Base pairs usually form in a nested fashion in RNA secondary structures, similar to the pairing of left and right parentheses in a mathematical expression. The exception to the rule is an RNA pseudoknot, which is a formation having non-nested base pairing. An RNA pseudoknot may be mingled with another RNA tertiary motif [215], and found in non-coding RNAs [142, 176]. An RNA pseudoknot [8, 363], has important functional roles in many biological processes [307, 345]. For instance, an RNA pseudoknot is required for telomerase activity [363]. An RNA pseudoknot has also been shown to regulate ribosomal frameshifting efficiency in a virus [276].

Detection, evaluation and analysis of RNA pseudoknots represent active areas of research. In particular, pseudoknot alignment between a pair of structures or among multiple structures is a focus of study [259, 320, 344, 407]. This case study presents an approach, called RKalign, for RNA pseudoknot alignment [340]. RKalign compares two pseudoknotted RNAs represented by sequence data (i.e., nucleotides or bases) and corresponding structure data (i.e., base pairs). The structure data of a pseudoknotted RNA is obtained from literature or from public databases [18, 38, 361, 370]. Given two or more pseudoknotted RNAs, an alignment of the RNA molecules can be computed.

RKalign adopts the partition function methodology to calculate the posterior probabilities or log-odds scores of structural alignments. Partition function computations have been used to calculate the probabilities of particular base pairings or structures in the study of RNA molecule folding [104, 255]. The partition function provides a confidence estimate in the quotient of the sum of all structures containing a specific base pair divided by the sum over all structures. RKalign demonstrates that the partition function approach can be applied to sequence and structure alignments.

Posterior probabilities are used to compute an alignment of biomolecules [263, 311]. In these cases, the partition function was employed to calculate the posterior probabilities of protein sequence alignments. A similar technique was developed using hidden Markov models (HMMs) to calculate the posterior probabilities [102]. These approaches have been further extended to structure-based multiple RNA alignments where partition functions were calculated inside and outside of subsequence pairs within two pseudoknot-free RNA sequences [395]. In this section, the partition function methodology is applied to alignments of RNAs with pseudoknots.

Several tools are available for RNA sequence-structure alignment [252, 367, 394]. Most RNA sequence-structure alignment tools do not consider a pseudoknot in the structure and assume that a structure is pseudoknot-free. One method by Mohl, et al., [264] performs sequence-structure alignment for RNA pseudoknots which involves a pipeline for combining alignment and prediction of pseudoknots. Han, et al., [155] decomposed embedded pseudoknots into simple pseudoknots and aligned them recursively. Yoon [412] used a profile HMM (hidden Markov model) to establish sequence alignment constraints, and incorporated these constraints into an algorithm for aligning RNAs with pseudoknots. Wong, et al., [400] identified the pseudoknot type of a given structure and developed dynamic programming algorithms for structural alignments of different pseudoknot types. Finally, Huang, et al., [176] applied a tree decomposition algorithm to search for non-coding RNA pseudoknot structures in genomes.

The above pseudoknot-aware methods align a pseudoknot structure with an RNA sequence or genome. During the alignment process, the RNA sequence is folded and its secondary structure is predicted. Xu, et al., [407] presented a different method, called RNA Sampler, which simultaneously folds and aligns multiple RNA sequences while taking into account a possible pseudoknot. Similar techniques are implemented in DAFS [320] and SimulFold [259]. Additional methods can be found in the CompaRNA Web server [300].

While the methods mentioned above perform alignment and folding simultaneously, RKalign aligns two RNA molecules whose structures, including possible pseudoknots, are known. RNA molecules known to contain pseudoknots can be found in publicly accessible databases [18, 38, 370]. RNA pseudoknot researchers continually add and annotate novel pseudoknot structure information to these publicly accessible databases. Tools like RKalign perform invaluable analysis of data in these repositories.

There are two groups of algorithms capable of aligning known RNA structures. The first group aligns two RNA three-dimensional (3D) structures containing pseudoknots. One algorithm takes into account nucleotide, dihedral angle and base-pairing similarities between structures [118]. Another aligns two RNA 3D structures based on a unit-vector root-mean-square approach [66]. Others employ a structural symbolic alphabet of different nucleotide conformations to align RNA 3D structures [74, 377]. One pairwise comparison method is based on the 3D similarity of generalized secondary structure units [171]. The R3D Align tool executes global pairwise alignment of RNA 3D structures using local superpositions [303]. He, et al., developed the RASS Web server that compares RNA 3D structures by considering both sequence and 3D structure information [160]. The above methods and tools align a pair of RNA tertiary structures by considering their geometric properties and/or torsion angles.

The second group of algorithms is concerned with aligning two RNA secondary structures without pseudoknots. Some of these algorithms employ general edit-distance alignment [419] or tree matching techniques [168, 238, 331]. Another approximates an alignment between a pseudoknot-free structure and a pseudoknotted structure [185].

RKalign is unlike the methods above in that it computes an alignment between two RNA secondary structures known to contain pseudoknots by utilizing the partition function methodology [340, 341]. One method similar to RKalign is the CARNA tool [344]. Like RKalign, CARNA aligns two known RNA secondary structures containing pseudoknots and employs constraint programming techniques with an iterative branch and bound scheme. In this case study, extensive experiments compared RKalign with CARNA using multiple different datasets for benchmarking.

4.5.2 Methods

Figure 4.8 illustrates a structural alignment by RKalign [340, 341] of two pseudoknotted RNA molecules from the PDB [38] crystal structure database. PDB molecules 1E8O [388] and 2B57 [137] are shown in raw FASTA format under the word "INPUT." In each molecule, stem base pairings are depicted by parentheses and brackets. The pseudoknotted nature of each PDB molecule is evidenced by the non-nested condition of the base pairs. Note that pairs of parentheses characters are all nested relative to other parentheses characters, and that pairs of bracket characters are all nested relative to other bracket characters. However, parentheses and square bracket characters are not nested relative to each other.

Under "OUTPUT" in Figure 4.8 is "Runtime: 172 ms," which indicates that RKalign structurally aligned PDB molecules 1E8O and 2B57 in 172 milliseconds. The Alignment computed by RKalign is shown in the figure below the run time information. Good structural alignment occurs when base pairs in the first structure are aligned with base pairs of the second structure. A comparison of RKalign with other structural alignment tools is detailed below.

The time complexity of the RKalign pairwise alignment algorithm is $O(m^2n^2)$ where m and n are the lengths of the two sequences aligned [340, 341]. As part of the RKalign alignment process, a two-dimensional matrix is maintained in storage and requires $O(mn)$ space. Thus, the space complexity of the algorithm is $O(mn)$. The time complexity is "worst case" since in calculating the partition functions, some base pairs are ignored. During experiments described here, over 200 alignments are evaluated. Experiment running times range from 16 ms to roughly 7 minutes and lengths of the aligned structures range from 22 to 1,553 nt.

```
INPUT

>1E8O
gGGCCGGGCGCGGUGGCGCGCGCCUGUAGUCCCAGCUACUCGGGAGGCUC
(((((((((((((..[[))))))))....((((.]].....))))))))

>2B57
GACAUAUAAUCGCGUGGAUAUGGCACGCAAGUUUCUACCGGGCACCGUAAAUGUCCGAUUAUGUC
((((((.....(((((.[[[.[[))))).........(((((((]]..]]]))))))..))))))

OUTPUT

RKalign structure alignment

Runtime: 172ms

Alignment:

>1E8O
gGGCC------GGGCGCGGU---GGCGCGCGCC--UG-U---AGUC--CCAGCUAC---UCGG--GA---GGCUC
(((((------((((((((.----.[[))))))--).-.----..((--((.]]......--..))--))---)))))

>2B57
GACAUAUAAUC-G-CGUG-GAUAUGG-CACG-CAA-GUUUCUACCGGGCA-CCG-UAAA--UGUCCGAUUAUGUC
((((((.....-(-(((((-.[[[.[[-))))-)..-........(((((((-]].-.]]]--))))))..))))))
```

FIGURE 4.8: PDB crystal structures 1E8O and 2B57. Each contains a pseudoknot shown before and after being structurally aligned by RKalign.

RNA pseudoknot method comparison experiments: pairwise alignment

RKalign is implemented in Java. The program accepts as input two pseudoknotted RNAs. Each RNA has both sequence data (i.e., nucleotides or bases) and structure data (i.e., base pairs), and produces as output an alignment between the two pseudoknotted RNAs. Popular benchmark datasets such as BRAliBase [131], RNase P [56] and Rfam [61] are not suitable for testing RKalign. This is because BRAliBase contains only sequence information and RNase P and Rfam contain consensus structures of multiple sequence alignments rather than alignments of individual structures of RNAs. As a consequence, two datasets were manually created for testing RKalign and comparing it with related alignment methods. The first denoted Dataset1 contains 38 RNA pseudoknot structures chosen from the PDB [38] and RNA STRAND [18]. These RNAs were selected in such a way that they have a wide range of sequence lengths. Each three-dimensional (3D) molecule in this dataset was taken from the PDB. The secondary structure of the 3D molecule was obtained with RNAview [408], retrieved from RNA STRAND. Dataset2 contains 36 RNA pseudoknot structures chosen from PseudoBase [361, 370]. As in the first dataset, the RNA molecules in the second dataset have a wide range of sequence lengths. The pseudoknots can be broadly classified as H-type and recursive pseudoknots [345, 400]. There are 12 H-type pseudoknots and 26 recursive pseudoknots in Dataset1 and 22 H-type pseudoknots and 14 recursive pseudoknots in Dataset2.

Alignment quality

A good structural alignment, as mentioned above, tends to align a base pair with another base pair rather than with two single bases [171, 303]. Therefore, the base_mismatch ratio is used to assess the quality of an alignment. A base mismatch occurs when a single base is aligned with the left or right base of a base pair or when a nucleotide is aligned to a gap. The base_mismatch ratio (BMR) of an alignment $a_{A,B}$ between structure A and structure B is defined as:

$$BMR(a_{A,B}) = \left(\frac{base\_mismatches(a_{A,B})}{alignment\_lines(a_{A,B})} \right) \times 100\%, \qquad (4.6)$$

i.e., $BMR(a_{A,B})$ is the number of base_mismatches in $a_{A,B}$ divided by the total number of alignment lines in $a_{A,B}$, multiplied by 100%. Statistically significant performance differences between alignment methods are calculated using Wilcoxon signed rank tests [393], which are commonly used for comparing alignment programs [277, 364, 397]. Any p-value below 0.05 is considered statistically significant [277, 364, 397].

4.5.3   Results

Experiments were conducted to evaluate the performance of RKalign and compare it with related methods. The base_mismatch ratio was used as the performance measure. In the first experiment, 106 pairs of RNA pseudoknot structures were selected from Dataset1. The method consisted of aligning the two molecules in each of the 106 pairs. The two molecules in a pair belonged to the same pseudoknot type because there is little value in aligning RNA molecules that lack consensus [53, 171]. The average base_mismatch ratio calculated by RKalign for the selected 106 pairs was 34.84%, compared to the average base_mismatch ratio, 78.53%, for all pairs of molecules in Dataset1.

In addition, other algorithms run were CARNA [344], RNA Sampler [407], DAFS [320], R3D Align [304] and RASS [160] on the 106 pairs of molecules. CARNA was chosen because an option of the tool is closely related to RKalign, both of which can align known pseudoknot structures.

RNA Sampler and DAFS were chosen because they are widely used tools capable of simultaneously folding and aligning RNA sequences on pseudoknots without known structures. When running these two tools, the structure information in Dataset1 was ignored and only the sequence data was used as the input of the tools.

R3D Align and RASS were chosen because they are state-of-the-art RNA 3D alignment programs; furthermore, like RKalign, R3D Align and RASS output the entire alignments of two RNA structures. Since R3D Align and RASS accept 3D structures as input whereas RKalign and CARNA accept bases and base pairs as input, the PDB files in Dataset1 were used as the input for R3D Align and RASS. Corresponding RNA STRAND entries from Dataset1 were used as the input for RKalign and CARNA. Figure 4.9 presents histograms for the base_mismatch ratios of the six tools. Figure 4.10 presents boxplots for the base_mismatch ratios of the six tools. These figures show the distribution of the base_mismatch ratios. RKalign and CARNA were not statistically different according to a Wilcoxon signed rank test ($p > 0.05$). On the other hand, they both performed significantly better than the other four tools according to the Wilcoxon signed rank test ($p < 0.05$).

It was observed that the structures predicted by RNA Sampler and DAFS might not be correct. Consequently, there were many base mismatches with respect to the known structures in the alignments. For example, consider Figure 4.11, which shows the alignment results of DAFS, R3D Align and RKalign on two pseudoknot structures with PDB IDs 1L2X [109] and 1RNK [335]. The base_mismatch ratios of DAFS (R3D Align, RKalign, respectively) were 57.14% (67.39%, 27.78%, respectively). Figure 4.11(a) shows the predicted common secondary structure and the alignment produced by DAFS. Figure 4.11(b) shows the known secondary structures of 1L2X and 1RNK and the alignment produced by DAFS where the known secondary structures are used to calculate the base_mismatch ratios. Figure 4.11(c) shows the alignment obtained from R3D Align and Figure 4.11(d) shows the alignment obtained from RKalign. It can be seen that the predicted common secondary structure in Figure 4.11(a) is quite different from the known secondary structure of 1L2X. Refer to Figure 4.11(b). The base G (G, C, C, A, A and A, respectively) at position 1 (2, 8, 22, 23, 24 and 25, respectively) in 1L2X is a single base, which is aligned with the left or right base of some base pair in 1RNK, leading to base mismatches in the alignment. Similarly, the base G (A, C, A and U respectively) at position 7 (20, 21, 24 and 34, respectively) in 1RNK is a single base, which is aligned with the left or right base of some base pair in 1L2X. R3D Align did not align the pseudoknot structures well either because many gaps are involved in the alignment (Figure 4.11(c)). In this example, RKalign produced the best alignment (Figure 4.11(d)). It should be pointed out, however, that 3D alignment programs such as R3D Align are general purpose structure alignment tools capable of comparing two RNA 3D molecules with diverse tertiary motifs, whereas RKalign focuses on secondary structures with pseudoknots only.

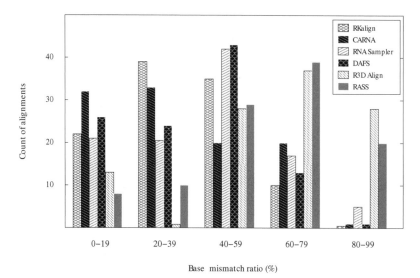

FIGURE 4.9: Histograms for the base_mismatch ratios of the alignments produced by RKalign, CARNA, RNA Sampler, DAFS, R3D Align and RASS, respectively, on the 106 structure pairs selected from Dataset1. Buckets on the x-axis are defined by equal-width ranges 0 to19, 20 to 39, 40 to 59, 60 to 79, and 80 to 99 (rounded down to the nearest whole number). These histograms show the distribution of the base_mismatch ratios of the alignments produced by the six tools.

In a separate experiment, RKalign, CARNA, RNA Sampler and DAFS were compared using the RNA structures in Dataset2. One hundred twenty four pairs of molecules from Dataset2 were selected. In each pair selected, both molecules belonged to the same pseudoknot type. The average base_mismatch ratio calculated by RKalign for the selected 124 pairs was 35.89%, compared to the average base_mismatch ratio, 81.56%, for all pairs of molecules in Dataset2. Each of the four tools was applied to the molecules to produce 124 pairwise alignments.

Figure 4.12 presents histograms for the base_mismatch ratios of the four tools. Figure 4.13 presents boxplots for the base_mismatch ratios of the four tools. These figures show the distribution of the base_mismatch ratios. RKalign and CARNA were not statistically different (Wilcoxon

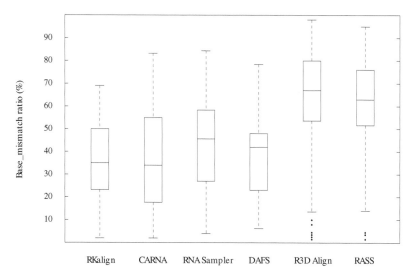

FIGURE 4.10: Boxplots for the base-mismatch ratios of the alignments produced by RKalign, CARNA, RNA Sampler, DAFS, R3D Align and RASS, respectively, on the 106 structure pairs selected from Dataset1. The median of the base-mismatch ratios yielded by RKalign (CARNA, RNA Sampler, DAFS, R3D Align, RASS, respectively) is 35.29% (34.38%, 45.86%, 41.99%, 67.57%, 63.29%, respectively).

signed rank test, $p > 0.05$); both tools performed significantly better than RNA Sampler and DAFS (Wilcoxon signed rank test, $p < 0.05$). Based on the above results, there is no statistically significant difference between RKalign and CARNA in terms of base-mismatch ratios. Note that a good pseudoknot alignment has many matched stems and few mismatched stems [176, 342].

As a final case study experiment, RKalign was compared with CARNA by examining how they match stems in pseudoknot structures A and B. A stem $s_A \in A$ is said to match a stem $s_B \in B$ if:

- s_A and s_B are aligned together and cannot be aligned with other stems.

- For every base pair $x \in s_A$ and base pair $y \in s_B$, a base of x is aligned with a base of y if and only if the other base of x is aligned with the other base of y.

Otherwise, there is a stem mismatch between s_A and s_B. The stem-mismatch ratio of an alignment $a_{A,B}$ between structure A and structure B is defined as $(1 - M)$ where M is the number of matched

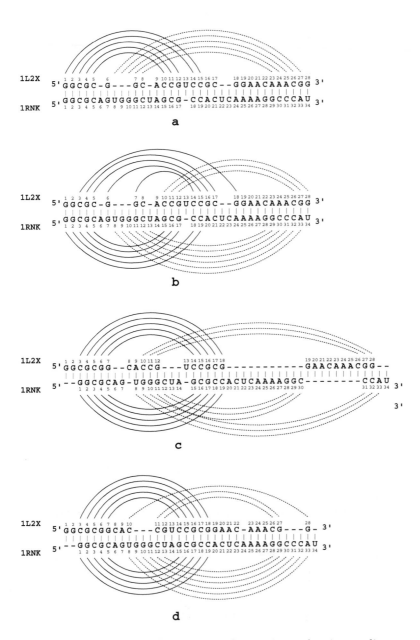

FIGURE 4.11: Example showing base mismatches in an alignment produced by DAFS, R3D Align, and RKalign.

stems in $a_{A,B}$ divided by the total number of stems in A and B, multiplied by 100%. Figure 4.14 shows the average stem-mismatch ratios of RKalign and CARNA obtained by running the tools on Dataset1 and Dataset2. RKalign performed significantly better than CARNA (Wilcoxon signed rank test, $p < 0.05$).

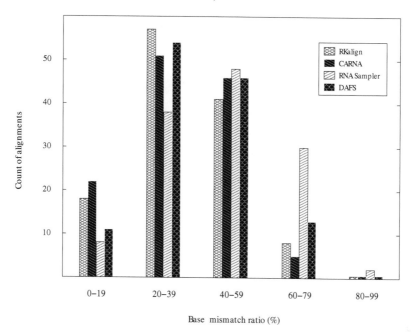

FIGURE 4.12: Histograms for the base-mismatch ratios of the alignments produced by RKalign, CARNA, RNA Sampler and DAFS on 124 structure pairs selected from Dataset2. Buckets on the x-axis are defined by equal-width ranges 0 to19, 20 to 39, 40 to 59, 60 to 79, and 80 to 99 (rounded down to the nearest whole number). These histograms show the distribution of the base-mismatch ratios of the alignments produced by the four tools.

4.5.4   Discussion and Conclusions

The RKakign method adopts the partition function methodology to calculate the posterior log-odds scores of the alignments between bases or base pairs of the RNAs with a dynamic programming algorithm [340, 341]. The posterior log-odds scores were used to calculate the expected accuracy of an alignment between the RNAs. The goal was to find an optimal alignment with the maximum expected accuracy and a

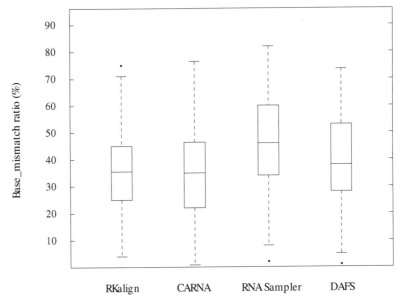

FIGURE 4.13: Boxplots for the base_mismatch ratios of the alignments produced by RKalign, CARNA, RNA Sampler and DAFS on 124 structure pairs selected from Dataset2. The median of the base_mismatch ratios yielded by RKalign (CARNA, RNA Sampler, DAFS, respectively) is 36.31% (35.81%, 45.83%, 39.29%, respectively).

heuristic was devised to achieve this goal. Experimental results demonstrated the good performance of the RKalign method. New pseudoknotted structures are found periodically, as exemplified by the recently determined ribosomal CCR5 frameshift pseudoknot [35] and the translational enhancer structures found in the 3′ untranslated regions (UTRs) of plant viruses [127, 128, 256, 353]. It is important to be able to compare these new structures to a database of known pseudoknots to determine the possibility of similar functionality. For example, recently discovered pseudoknots found in the 3′ UTRs of plant viruses have been shown to act as translational enhancers and have 3D structures similar to tRNAs. It is significant that these RNA structures contain pseudoknots that produce tRNA-like 3D folds, but are not derived from the standard tRNA secondary structure cloverleaf. In addition, these elements have been shown to be important for ribosome binding. RKalign is a valuable tool when searching databases for structure-function analysis of pseudoknots.

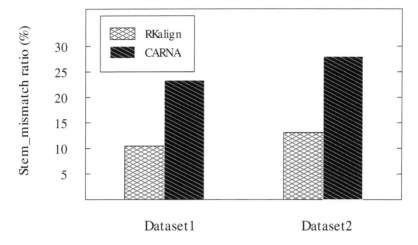

FIGURE 4.14: Comparison of the stem_mismatch ratios yielded by RKalign and CARNA. Average stem_mismatch ratios of the alignments produced by RKalign and CARNA on the 106 structure pairs selected from Dataset1 and the 124 structure pairs selected from Dataset2, respectively. For Dataset1, the average stem_mismatch ratio of RKalign is 10.8% and the average stem_mismatch ratio of CARNA is 23.5%. For Dataset2, the average stem_mismatch ratio of RKalign is 13.1% and the average stem_mismatch ratio of CARNA is 28.9%. RKalign performed significantly better than CARNA in terms of stem_mismatch ratios (Wilcoxon signed rank test, $p < 0.05$).

RNA pseudoknot method comparison experiments: multiple alignment

The RKalign pairwise alignment method can be extended to align multiple RNA pseudoknot structures by utilizing a guide tree [340, 341]. Specifically, each structure is treated as a cluster and the expected accuracy [104] is used as the measure to determine the similarity of two structures or clusters. Initially, two RNA structures that are most similar were merged into one cluster. Subsequently, two clusters that were most similar were merged into a larger cluster using the agglomerative hierarchical clustering algorithm [157], where the similarity of two clusters is calculated by the average linkage algorithm. An alignment of two clusters is actually an alignment of two profiles, where each cluster is treated as a profile. Initially, each profile contains a single RNA pseudoknot structure. As the guide tree grows, a profile may contain multiple RNA pseudoknot structures; more precisely, the profile is a multiple alignment of these RNA structures. A single base of a profile is a column of the profile where the column contains single bases or gaps; a base pair of a profile includes two columns of the profile where the left column contains left bases or gaps and the right column contains corresponding right bases or gaps, and left bases and corresponding right bases form base pairs.

Figure 4.15 illustrates a multiple alignment by RKalign [340, 341] of six pseudoknotted RNA molecules from the PDB crystal structure database [38]. PDB molecules 1E8O [388], 1XP7 [112], 2F4X [204], 2OOM [223], 2D19 [27] and 1BAU [268] are shown in raw FASTA format under "INPUT." In each molecule, stem base pairings are depicted by parentheses and brackets. As previously mentioned, the pseudoknotted nature of each PDB molecule is evidenced by the non-nested condition of the base pairs. Note that pairs of parentheses characters are all nested relative to other parentheses characters, and that pairs of bracket characters are all nested relative to other bracket characters. However, parentheses and square bracket characters are not nested relative to each other.

Under "OUTPUT" in Figure 4.15, "Runtime: 625 ms" indicates that RKalign structurally aligned the six PDB molecules in 625 milliseconds. The alignment computed by RKalign is shown in the figure below the run time information.

The time complexity of the multiple alignment algorithm is $O(k^2 n^4)$, where k is the number of structures in the alignment and n is the maximum of the lengths of the structures; the space complexity of the algorithm is $O(k^2 n^2)$. The algorithm was tested by selecting 30 groups each having 3, 4, or 5 pseudoknot structures of the same type from the datasets used in this study, and by performing multiple alignment in each group. This algorithm was compared with three related methods: CARNA [344], RNA Sampler [407] and DAFS [320]. The base_mismatch ratio (BMR) of a multiple alignment MA was defined as the sum of base_mismatch ratios of all pairs of structures in MA divided by the total number of structure pairs in MA, multiplied by 100%. The average base_mismatch ratio of RKalign (CARNA, RNA Sampler, DAFS, respectively) was 26.01% (25.79%, 32.15%, 29.23%, respectively). RKalign and CARNA were not statistically different (Wilcoxon signed rank test, $p > 0.05$); the two methods were significantly better than RNA Sampler and DAFS (Wilcoxon signed rank test, $p < 0.05$).

Both pairwise alignment and multiple alignment programs are available in the RKalign tool. RKalign is capable of accepting as input pseudoknotted RNAs with both sequence (nucleotides or bases) and structure data (base pairs), and producing as output an alignment between the pseudoknotted RNAs. As more pseudoknots are revealed, collected

```
INPUT

>1E8O
gGGCCGGGCGCGGUGGCGCGCGCCUGUAGUCCCAGCUACUCGGGAGGCUC
((((((((((((((..[[))))))))))....((((.]]....)))))))))

>1XP7
CUUGCUGAAGUGCACACAGCAAGCUUGCUGAAGUGCACACAGCAAG
((((((((..[[[[[[.)))))))((((((((..]]]]]].)))))))

>2F4X
GGUUGCUGAAGCGCGCACGGCAACGGUUGCUGAAGCGCGCACGGCAAC
.(((((((..[[[[[[.))))))).((((((((..]]]]]].)))))))

>2OOM
GGAGCCUGGGAGCUCCCACGGUCCcaGACGUG
(((((([[[[[.)))))((((((.]]]]])))))

>2D19
GCUGAAGUGCACACGGCGCUGAAGUGCACACGGC
((((..[[[[[[.))))((((..]]]]]].))))

>1BAU
GGCAAUGAAGCGCGCACGUUGCCGGCAAUGAAGCGCGCACGUUGCC
(((((((.[[..[[[[)))))))((((((((.]]]]..]]))))))))

OUTPUT

RKalign structure alignment

Runtime: 625ms

>1E8O
gGGCCG-GGC-GCGGU---G--GC---GCGC-GCCUGUAGU-C-C-C-A---GC-UACU-C---G-G-G-AGGCUC
((((((-(((-(((((.---.--[[---))))-))))....-(-(-(-(-.---]]-....-.---)-)-)-)))))))

>1XP7
-------CUUGCUG-A---AGUGCACA-CAGCAAG-----CUUGCUG-AAGUGCA----CA---CAGCAAG-----
-------((((((((-.---.[[[[[[.-)))))))-----((((((((-..]]]]]----].---)))))))-----

>2F4X
------GGUUGCUG-A---AGCGCGCA-CGGCAAC-G---GUUGCUG-AAGCGCG----CA---CGGCAAC-----
------.((((((((-.---.[[[[[[.-))))))-.----((((((((-..]]]]]----].---)))))))-----

>2OOM
-------GG-AGC----C---UGGG-A--GCU-CC-------CA-CGGU--CCca------GA-CG-UG-------
-------((-(((----[---[[[-.--)))-))-------((-(((.--]]]]------])-))-)-)-------

>2D19
----------GCUG-A---AGUGCACA-CGGC-----------GCUG-AAGUGCA----CA---CGGC--------
----------((((-.---.[[[[[[.-))))-----------((((-..]]]]]----].---))))--------

>1BAU
-------GGCAAUG-AAGCG-CGCA---CGUUGCC-----GGCAAUG-A--AGCG--C--GC-ACGUUGCC-----
-------((((((((-.[[..-[[[[---)))))))-----((((((((-.--]]]--.---.]-)))))))-----
```

FIGURE 4.15: Multiple alignment: PDB crystal structures 1E8O, 1XP7, 2F4X, 2OOM, 2D19 and 1BAU, each containing a pseudoknot, are shown before and after being structurally aligned by RKalign.

and stored in public databases, a tool like RKalign will play a significant role in data comparison, annotation, analysis, and retrieval in these databases. RKalign is designed to align known RNA pseudoknot structures. A different approach is to simultaneously fold and align RNA sequences without known structures, as adopted by several existing tools [259, 320, 407]. When the structure information is not available, this simultaneous folding and alignment approach is the best. However, when pseudoknot structures already exist, RKalign performs significantly better than the existing tools, as observed in the experiments. The reason is that the structures predicted by these tools may not be correct. As a consequence, there are many base mismatches with respect to the known structures in the resulting alignments. A pseudoknot is one type of RNA tertiary motif [405]. There are 3D alignment programs that can compare RNA tertiary structures including pseudoknots [160, 303]. These programs consider the entire RNA 3D structure as a whole and accept PDB files with 3D coordinates as input.

In the experiments aligning secondary structures with pseudoknots, RKalign outperforms current 3D alignment programs. This is expected since current alignment programs are general-purpose structure alignment tools capable of comparing two RNA 3D molecules with diverse tertiary motifs, whereas RKalign deals with secondary structures with pseudoknots only.

The experiments described here focus on pseudoknot alignment using RKalign. However, RKalign also aligns RNA secondary structures without pseudoknots. RKalign aligned 102 pairs of pseudoknot-free structures taken from RNA STRAND where the pseudoknot-free structures belonged to Rfam [61]. RKalign was compared with three other tools: CARNA [344], RNAforester [168] and RSmatch [238]. RNAforester, included in the widely used Vienna RNA package [169], is a versatile RNA structure alignment tool. Like RKalign and CARNA, an option of RNAforester is able to accept as input two RNA molecules with both sequence data (nucleotides or bases) and secondary structure data (base pairs), and produce as output the global alignment of the two molecules. However, a limitation of RNAforester is that the aligned secondary structures cannot contain pseudoknots. RSmatch is similar to RNAforester, sharing the same limitation. Experimental results showed that the average base_mismatch ratio for RKalign (CARNA, RNAforester, RSmatch, respectively) was 43.52% (42.27%, 35.11%, 39.66%, respectively), indicating RNAforester performed best. These results were expected since RKalign is designed for comparing complex pseudoknot structures whereas RNAforester focuses on simpler pseudoknot-free structures.

The tool most closely related to RKalign is CARNA [344]. Both methods are able to accept as inputs known pseudoknot structures and produce as outputs alignments of the known structures. Experimental results indicated that the two methods perform well in terms of base_mismatch ratios, although RKalign yields much lower stem_mismatch ratios. In the experiments described here, the input data of RKalign are fixed structures, i.e., RNA sequence and structure data. The main purpose of CARNA is to align dot plots [170], and its scoring is optimized for the dot plot graphical data format. Therefore, CARNA is preferred for aligning dot plots and RKalign is preferred for aligning fixed structures.

CHAPTER 5

Biological Data Mining

CONTENTS

5.1   INTRODUCTION

Previous chapters describe the storage, cleaning, integration and searching of biological data. In this chapter, the extraction of useful information from biological data repositories is explored. These repositories are often quite large. Extraction of valuable "nuggets of information" from large collections of data is known as data mining. Gathering of such useful information from collections of biological data is simply called biological data mining.

Electronic biological data are accumulating in databases at exponential rates with no end in sight. The challenge presented to biological data mining professionals is to develop efficient and cost-effective algorithms to locate or predict the location of information in these massive collections. High throughput parallel sequencing technologies allow hundreds of millions of DNA molecules to be sequenced at the same time and at ever-increasing speeds [79]. Simultaneously, new parallel computational approaches for analyzing this data, including CUDA [317] and Apache Hadoop (http://hadoop.apache.org/), help to provide essential knowledge about the origins of a variety of common diseases.

By combining massively parallel processing methodologies with disciplined data mining methodologies, bioinformatics researchers are better equipped to solve the mysteries of disease. Common data mining tasks include association rule learning, clustering, classification, regression, summarization and sequential pattern mining.

In the field of bioinformatics, clustering is invaluable for grouping sequences of DNA or non-coding RNA that have similarities. By using clustering techniques, scientists are able to better understand how various species are similar or how much two organisms that appear to be very different actually have in common. Non-coding RNA molecules are known to regulate gene expression and perform other vital functions in cells. By clustering non-coding RNA from many different disparate organisms, scientists may discover common ancestries that were previously unknown.

The clustering of gene sequences involves applying a range of standard distance measures and/or tree-construction algorithms to the collection of gene sequences available. Genes in each cluster have many similarities in comparison to one another but are dissimilar to genes in other clusters [201].

The classification of an unknown object into one of a finite set of categories helps the researcher to learn more about the unknown object. In bioinformatics, classification is valuable for predicting and/or confirming certain attributes of cellular entities.

This chapter introduces data mining in general (e.g., the background, definition, and process), describes data mining aspects of biological data mining and then presents two biological case studies related to biological data mining and biological pattern discovery. In both case studies, the targeted data mining information is non-coding RNA-related (i.e., ncRNA) [237]. ncRNA biological data mining presents unique challenges due to the nature of its configuration at rest (at minimal thermal energy state). ncRNA exists in various forms and serves a variety of critical cellular functions.

5.2 GENERAL DATA MINING

While data mining mainly originated from work in the field of statistics and machine learning as an interdisciplinary field, it advanced to include pattern discovery, database analysis, results visualization, and other tasks.

The use of statistics has been the principle data analysis method used in most scientific fields in recent history [411]. Data mining is the analysis of (often large) observational data sets to find unsuspected relationships (i.e., golden nuggets of information) and to summarize the data in novel ways that are understandable and useful [156, 158, 358]. Although data mining is fundamentally based on mathematics, specifically statistics, many data mining approaches partially adopt heuristics (in addition to mathematics) to resolve real-world problems, especially when categorical (discrete) data are used.

According to Cross-Industry Standard Process for Data Mining (CRISP-DM) [334], the data mining process consists of six steps: business understanding, data understanding, data preparation, modeling, evaluation, and deployment.

While statistics uses a sample of data drawn from a population, data mining typically uses data encompassing an entire population. In statistics, the use of a sample of data provides an almost identical statistical significance as would occur if the researcher used population data. Sampling probably originated as a means to test large data sets several decades ago when computer systems or manual calculations could not handle an entire population data sample. Conversely, because data mining clusters data and discovers hidden patterns, it uses the entirety of the population data. The exception to this rule is the occasional use of a data mining approach that may partially or temporarily use a sample of data to address scalability issues.

As described in a previous chapter, the quality of data in the biomedical and health care fields often needs improvement [411]. Medical data often contains missing values because patients with similar diseases do not always undergo the same examinations and laboratory testing. This may generate different data sets. In addition, medical data often contains time-series attributes. Researchers must handle these data sets with special consideration of the time element.

Generally, data mining algorithms are classified as descriptive (or unsupervised learning) and predictive (or supervised learning). Descriptive data mining clusters data by measuring the similarities between objects (or records) and discovers unknown relationships in data so that users can readily understand huge amounts of information. Descriptive data mining is exploratory in nature and includes clustering, association and summarization. Predictive data mining infers prediction rules (a.k.a. classification and prediction models) from (training) data and applies the rules to unseen (i.e., testing) data. Predictive data mining includes classification, regression and prediction.

Data mining faces several challenges that hamper its clinical usage by health professionals.

- Data mining algorithms usually require user-specified parameter(s) mainly because each data mining algorithm has its own theoretical assumptions. End users usually do not have sufficient information about the parameter(s) and their selection. To make the problem worse, data mining results are usually very sensitive to the parameter(s).

- Data mining accuracy is normally not adequate for use in a clinical environment. The generally low quality of patient data contributes to the problem because hospital information systems are usually designed for financial and billing purposes. Another reason for the low data quality is that biomedical and health care factors that affect diseases are not fully known.

- There is a shortage of full data mining packages for knowledge discovery.

An ideal data mining package should:

- Support intelligent data preprocessing that automatically selects and eliminates data for the purpose of mining and uses domain knowledge for various processes.

- Fully automate the knowledge discovery process so that it utilizes existing knowledge in data mining processes for better knowledge discovery.

To help researchers understand the importance of data mining, and application of data mining techniques, three of the most widely-used data mining algorithms (data classification, pattern discovery and data clustering) will be discussed along with guidelines for their respective use.

5.3 BIOLOGICAL DATA MINING

5.3.1 Gene Ontology as Enabling Tool for Mining

Recently, there has been much discussion on how to report biology knowledge in information systems. While it is difficult to persuade existing stakeholders to alter processes, there is general consensus about the need for a common, concise vocabulary of biological terms. An ontology provides a vocabulary for representing and communicating knowledge about a topic [43]. The Gene Ontology (GO) Consortium was formed in 1998 to establish and maintain an ontology of gene information. Over time, additional biology and medical ontology groups formed and today they are collectively called the Open Biomedical Ontologies (OBO) Group [338]. The OBO Foundry consists of ontologies in various stages of maturity. Among the more mature ontologies are the Cell Ontology (CL), Gene Ontology (GO), Foundational Model of Anatomy (FMA) and the Zebrafish Anatomical Ontology (ZAO).

In 2000, the GO consortium was a joint project of three model organism databases: FlyBase, Mouse Genome Informatics (MGI) and the *Saccharomyces* Genome Database (SGD). The goal of the consortium was to produce a precisely defined, structured, common, controlled vocabulary describing the roles of genes and gene products in every organism [26]. Within the GO, the following ontologies developed:

- Biological process ontology: a biological objective to which the gene or gene product contributes

- Molecular function ontology: the biochemical activity of a gene product

- Cellular component ontology: a place in a cell where a gene product is active

The GO (http://www.geneontology.org) is a community bioinformatics resource [42]. Each GO entry has a unique numeric identifier. Table 5.1 shows the status of GO as of September 2011. Table 5.2 lists 12 model organisms selected for targeted curation. Each organism is shown with the name of its respective database.

GO information exists as a publicly available flat file. The current version of the full GO dataset is located at

```
ftp://ftp.geneontology.org/pub/go/ontology/go.obo
```

Predetermined sets of GO terms, called GO Slims, are used to aggregate gene product information. GO slims may be created by users according to their needs, and may be specific to species or to particular areas of the gene ontology. Go Slims provided by the Gene Ontology Consortium (GOC) are described at

```
http://www.geneontology.org/GO.slims.shtml
```

TABLE 5.1: Status of Gene Ontology (GO)

Element Group	Element Count
Biological process terms	21,394
Molecular function terms	9,062
Cellular component terms	2,896
Species with annotations (including strains)	367,887
Total annotated gene products	11,855,555
Manually annotated gene products	437,164

Table 5.3 identifies GO Slims maintained by GOC curators and others. As the GO flat file evolves, the Go Slims are updated simultaneously. Users can create customized GO Slims using the OBO-Edit tool available at

http://oboedit.org/ .

OBO-Edit is an open source ontology editor written in Java. As an example, there is a GO Slim dataset for the *Saccharomyces cerevisiae* yeast genome. The current version of the GO Slim yeast dataset can be downloaded from

ftp://ftp.geneontology.org/pub/go/GO_slims/goslim_yeast.obo .

As of September 2014, the number of GO entries in the GO Slim yeast dataset was 169.

In addition to OBO-Edit, another valuable GO utility is the AmiGO Web-based tool that provides access to all terms and annotations in the GO database [42]. AmiGO users can browse the ontology terms and search the annotations. AmiGO is available at

http://amigo.geneontology.org .

Each entry is classified as a biological process (27,264), molecular function (10,733) or cellular component (3,753). See Table 5.1. An interesting possibility for the GO is to compare computational biology experiments with wet lab results. Another potential use is to assist in establishing

TABLE 5.2: Twelve model organisms selected for targeted curation and their respective databases

Species	Species Database
Arabidopsis thaliana	*Arabidopsis* Information Resource (TAIR)
Caenorhabditis elegans	WormBase
Danio rerio	Zebrafish Information Network (Zfin)
Dictyostelium discoideum	Dictybase (i.e., amoeba)
Drosophila melanogaster	FlyBase
Escherichia coli	EcoliHub
Gallus gallus	AgBase (i.e., chicken)
Homo sapiens	Human UniProtKB-Gene Ontology Annotation (UniProtKB-GOA) @EBI
Mus musculus	Mouse Genome Information (MGI)
Rattus norvegicus	Rat Genome Database (RGD)
Saccharomyces cerevisiae	*Saccharomyces* Genome Database (SGD)
Schizosaccharomyces pombe	GeneDB *S. Pombe* (i.e., yeast)

TABLE 5.3: List of GO Slims maintained by GOC as part of GO flat file

Name	Slim Developer
Generic GO Slim	GO Consortium
Plant Slim	*Arabidopsis* Information Resource
Candida albicans	*Candida* Genome Database (i.e., fungus)
Protein Information Resource Slim	Darren Natale, PIR
Schizosaccharomyces pombe Slim	Val Wood, PomBase (i.e., yeast)
Yeast Slim	*Saccharomyces* Genome Database
Aspergillus Slim	*Aspergillus* Genome Data (i.e., mold)
Metagenomics Slim	Jane Lomax and InterPro group

gene regulatory networks (GRNs). The GOSLIM dataset for yeast contains 24 cellular component codes, 44 molecular function codes and 101 biological process codes. To map genes to GO terms, a mapping dataset is provided by NCBI at

ftp://ftp.ncbi.nlm.nih.gov/gene/DATA/gene2go.gz   .

As of September 2014, this gene2go mapping file contained 1,630,771 unique gene-GO mappings. As an example of a successful application of GO, Khaladkar, et al., [201] used GO to classify the function of a cluster of ncRNA genes. Hypergeometric testing identifies whether a cluster of ncRNA genes has a better than average chance of possessing a specific GO trait [201]. This can be valuable when analyzing unknown ncRNA genes.

5.3.2 Biological Pattern Discovery

Pattern or motif discovery is the task of finding recurring patterns in biological data. Patterns can be sequential, mainly when discovered in DNA sequences. They can also be structural (e.g., when discovering RNA motifs). Finding common structural patterns aids understanding of the mechanism of action (e.g., post-transcriptional regulation). Unlike DNA motifs, which are sequentially conserved, RNA motifs exhibit conservation in structure, which may be common even if the sequences are different. Over the past few years, hundreds of algorithms have been developed to solve the sequential motif discovery problem, while less work has been done for the structural case.

Finding recurring patterns or motifs in biological data gives an indication of important functional or structural roles. Motifs can be sequential or structural and are represented as sequences when they display repeated patterns in biological sequences. Motifs are structural when they represent patterns of conserved base pairs (e.g., RNA secondary structures). Knowing structural motifs in RNA leads to a better understanding of the mechanisms of action. Unlike DNA motifs, which are sequentially conserved, RNA motifs may share a common structure even in the case of low sequence similarity.

The structural motif discovery problem should not be confused with two related problems: RNA structure prediction and RNA consensus structure prediction. The former is required to predict the secondary structure of a single RNA sequence. In general, the predicted structure minimizes the total free energy. The latter operation is required to find a list of base pairs that can simultaneously be formed in a set of related RNA sequences. In this case, it is generally assumed that sequences are evolutionarily related and share a similar overall fold. Evolutionary conservation information is utilized to improve the accuracy of structure prediction processes.

5.3.3 Biological Data Classification

Classification is used to assign data into predefined categorical class labels [156, 158]. Class is the attribute or feature in a data set, in which users are most interested. It is defined as the dependent variable in statistics. To classify data, a classification algorithm creates a model consisting of classification rules. For example, automobile insurers developed classification models to categorize drivers' applications as risky or safe. In the medical field, classification can help define diagnosis and prognosis based on symptoms, history of family illness and health conditions.

Classification is a two-step process consisting of training and testing. In the first step, a training model which consists of classifying rules is built. The training model is first "taught" the rules. When the training model is tested, i.e., presented with new data not from the training data, the model is able to make a decision based on what it has learned from the training data.

A hypothetical example of a simple banking classification rule to determine "LoanRisk" class is shown below:

IF MarriedApplicant = YES AND SteadyEmployment = YES
THEN
LoanRisk = LOW
ELSE
LoanRisk = HIGH
END-IF

Some classification rules use mathematical formulas to determine class labels. Classifying rules are not necessarily 100% true; generally, rules with 90 to 95% accuracy are regarded as solid. The accuracy of a classifier (or classification model) depends on the degree to which classifying rules are true.

The second step, testing, examines a classifier using testing data for accuracy. The class labels for the test data are known. The classifier is expected to predict the class label for each test case based on how it has been taught. Generally, the testing process is very simple and computationally inexpensive as compared to the training step, which may be complex and require considerable computational resources.

An interesting technique in classification is the ensemble approach. The rationale behind the approach is that multiple classifiers (or classification models) working together can yield better classification accuracy than the use of a single classifier. As a simple example, if Classifiers A, B, and C predict that a hard-to-classify patient (patient1) has a disease and Classifiers D and E predict that patient1 does not have that disease, by using a voting strategy, the ensemble predictor would predict that patient1 has the disease. In some cases, all classifiers may be assigned different weights and the final ensemble prediction would then be a weighted average of the classifier votes. The classification case study in this chapter explores the ensemble method in greater detail.

5.3.4 Biological Data Clustering

Clustering is unsupervised learning that occurs by observing only independent variables (unlike supervised learning both independent and dependent variables) [156, 158, 358]. In order words, unlike classification, clustering does not use "class". In fact, this is the main difference between classification and clustering. For this reason, clustering may be best used for studies of an exploratory nature, especially if those studies encompass large amounts of data, but very little is known about the data (such as the mass of data typically generated by microarray gene expression analysis).

Clustering is used to group objects into a specific number of clusters so that the objects within a cluster have high similarity values and objects from different clusters have low similarity. Similarities between two objects are measured using their attribute values. A very early application of clustering in biology was to cluster similar plants and animals to create taxonomies based on their attributes (such as the number of petals and the number of legs). A number of clustering algorithms have been introduced and used over the last few decades. These algorithms are mainly categorized as hierarchical and partitional. Each category will be further discussed below.

Hierarchical agglomerative clustering algorithms successively merge the most similar two groups of objects based on the pairwise distances between the two groups until a termination condition holds, so that objects are hierarchically grouped. For this reason, hierarchical algorithms can be categorized effectively according to the respective methods of calculating the similarity (or distance) between two groups of objects. In order words, this categorization is based on how the representative object of each group for similarity calculation is selected.

While hierarchical clustering is agglomerative, i.e., starting with atomic elements and aggregating them into clusters of increasing size, divisive clustering starts with a complete data set and subdivides the data set into smaller partitions. A divisive clustering algorithm iteratively performs these two operations: (1) deciding the best cluster to be split; (2) deciding the best way to split the selected cluster [321].

Unlike hierarchical clustering algorithms, partitional clustering algorithms require a user to input a parameter k, which is the number of clusters. Generally, partitional algorithms directly relocate objects to k clusters. Partitional algorithms are categorized according to how they relocate objects, how they select a cluster centroid (or representative) among objects within a (incomplete) cluster, and how they measure similarities between objects and cluster centroids. For example, k-means, the most widely-used partitional algorithm, first randomly selects k centroids (objects), and then decomposes objects into k disjoint groups by iteratively relocating objects based on the similarity between the centroids and the objects. In k-means, a cluster centroid is the mean value of objects in the cluster. In many cases, the cluster centroids are not actual cluster objects. Unlike k-means, k-medoids selects the nearest object to the mean value of objects in a cluster.

The major advantage of partitional clustering algorithms over other methods is their superior clustering accuracy as compared with hierarchal clustering algorithms resulting from their global optimization strategy (i.e., the recursive relocations of objects). In addition, partitional algorithms can handle large data sets which hierarchal algorithms cannot (i.e., better scalability) and can more quickly cluster data. In short, partitional algorithms are more effective and efficient than hierarchical algorithms. One major drawback of partitional algorithms is that their clustering results depend on the initial cluster centroids to some degree because the centroids are randomly selected. Thus, clustering results obtained are a little different each time a partitional algorithm runs. Such a process is known as non-deterministic. A deterministic process, by comparison, will yield the same result each and every time it is run.

5.4   CASE STUDY IN BIOLOGICAL MOTIF DISCOVERY

As an example of biological motif or pattern discovery, a non-coding RNA motif discovery case study is presented [62]. Evolutionarily conserved functional domains of non-coding RNA on chromosome X (*roX1*) of the fruit fly have been identified in eight *Drosophila* species. Interestingly, within the *roX1* RNAs of these same *Drosophila* species, conserved primary sequences were also found. Specifically, three repeats of the nucleotide sequence GUUNUACG were localized in the 3′ end of the predicted *roX1* RNAs for these eight *Drosophila* species. A covariance model (CM) was used to search for the characteristic features of *roX1* functional domains as a way to classify new examples of these structured RNAs in other *Drosophila* species. In spite of high levels of genomic sequencing activities worldwide, annotation of the *Drosophila* species is still incomplete. Much chromosome coordinate information remains unknown. Whole genomes of *Drosophila* were obtained and scanned to identify results in available annotated regions, i.e., chromosomes or scaffolds. Using known *roX1* examples for comparative support, it is believed to be possible to predict novel *roX1* functional domains accurately from sequence information alone. Annotating *roX1* on a genomic scale provides insight into evolutionary processes among various species. The results of this case study indicate that a CM search and classification process is effective in mining *roX1* RNA genes. Furthermore, due to the flexibility of the CM search methodology, this mining approach may very likely prove successful for similar searches in various other organisms.

5.4.1 Background

Non-coding RNAs (ncRNAs) are functional RNA transcripts that are not translated into protein (i.e., they are not messenger RNAs). Research has shown that ncRNAs perform a wide range of functions in cells [88, 106, 254, 351]. RNA on the X chromosome (*roX1*) plays an essential role in equalizing the level of transcription on the X chromosome in *Drosophila* males and females. Like humans, the *Drosophila* male has a single X chromosome while the *Drosophila* female has two X chromosomes [290]. Experiments confirmed that *roX1* RNA exists in eight *Drosophila* species [287, 288, 289]. It is believed that there may exist secondary structure conservation of the *roX1* gene among other *Drosophila* species [287, 288]. Advances in the research of genomes from twelve *Drosophila* species [346] may shed light on this interesting issue.

A highly regarded covariance model (CM) method named Infernal has been successfully used in the classification of ncRNAs. Infernal is considered by many bioinformatics researchers to be one of the most accurate tools for this purpose [336, 376]. Infernal is a genome-wide search tool that applies stochastic context-free grammars expressed as CMs to find genomic regions that may contain ncRNAs [62, 355, 399]. A CM is a statistical representation of a group of RNAs that share a common consensus secondary structure [104]. The Infernal software package [142, 143, 270] contains a number of powerful utilities. One named cmbuild creates a CM from a Stockholm alignment of sequences. Another Infernal utility named cmsearch is used to scan a genome for sequences that match to the model and classify candidates as likely or unlikely to belong to the group the CM represents. The time to run a cmsearch process can be lengthy, depending on the size of the genome scanned. However, by utilizing parallel processing methods, results can be obtained in considerably shorter timeframes.

The first step in this case study was to demonstrate the capability of using a CM in a genome-scale homology search. Known *Drosophila* *roX1* sequences from eight species were gathered. These eight species are *D. ananassae*, *D. erecta*, *D. melanogaster*, *D. mojavensis*, *D. pseudoobscura*, *D. simulans*, *D. virilis* and *D. yakuba* [288]. Using a leave-one-out testing approach, covariance models were created using seven species at a time and the genome of the eighth species was scanned for a match. This demonstrated that the CM classification process is feasible in that six of the eight searches were successful. The next step was to scan the genomes of *Drosophila* species for which there were no confirmed *roX1* sequences. These four *Drosophila* species were *D. grimshawi*, *D. persimilis*, *D. sechellia*, and *D. willistoni*. Such a comparative genomics approach has been successful in the unicellular *Saccharomyces cerevisiae* [197], i.e., a species of yeast very commonly used in winemaking, baking and brewing.

Using this CM approach, the results show strong evidence of the presence of *roX1* functional domains in the genome of *D. sechellia*. This finding is believed to be novel and significant in ongoing genomic studies of *Drosophila* and related taxonomic groups. This bioinformatics study laid the groundwork for future CM ncRNA classification.

5.4.2 Methods

Eight *roX1* RNA sequences were obtained using wet lab experiments, i.e., confirmed versus predicted *in silico*. These sequences were from eight *Drosophila* species named *D. ananassae*, *D. erecta*, *D. melanogaster*, *D. mojavensis*, *D. pseudoobscura*, *D. simulans*, *D. virilis* and *D. yakuba* [288] (Table 5.4). In all eight cases, the sequences were expressed in standard FASTA format. The *roX1* RNA sequences are fairly large with lengths ranging from 3,433 to 3,768 nucleotides. The classification analysis dictated that the large sequences be subdivided into smaller subsequences while preserving all species and position information for each subsequence. The RSmatch software package used in this study has this sequence subdividing capability [238].

For tracking purposes, each of these eight *roX1* sequences processed by RSmatch's slide-and-fold function was assigned a unique five-character name. To illustrate with an example, one FASTA sequence named yp101 was annotated as: ">yp101 (1:3493) droana *roX1*". This name was formed as follows: in characters 1, 2 and 3: constant "yp1"; and in characters 4 and 5: a sequential two-digit number. Starting and ending positions for each of the eight sequences were described in FASTA notation as start:end where start represents the first numeric position and end represents the last numeric position. For each of the original sequences, "start" had the value of 1 and "end" had the value of the length of the sequence. The starting and ending positions for each sequence were formatted specifically for RSmatch [238]. When subsequences were extracted, the original FASTA notation was preserved and additional subsequence position information was inserted for sequence tracking purposes. Note that the length of the yp101 *roX1* gene sequence is 3,493 nucleotides (nt). RSmatch was used to extract 100 nt subsequences with 50 nt overlaps from this yp101 sequence. RSmatch produced properly annotated FASTA format sequences such as ">yp101:1-100 (1:3493) droana *roX1*", ">yp101:51-150 (1:3493) droana *roX1*", etc. This notation represents a 100 nt sequence extracted from positions 1 through 100 and positions 51 through 150 of the original yp101 sequence. All of the original FASTA notation information was retained in the FASTA notation of each subsequence. Providing position information in the notation of the extracted subsequences is a critical function performed by RSmatch.

In a similar manner, all eight *Drosophila roX1* sequences evaluated for this work were annotated for compatibility with RSmatch, thus preserving subsequence position information. RSmatch slide-and-fold was run with the following parameters: sequence size = 100 nt; overlap size = 50 nt; minimum free energy = 0. RNA structures were prepared using the slide-and-fold method. For each sequence, 100 nt subsequences were extracted at every 50 nt position from the 5′ end downstream to the 3′ end resulting in consecutive subsequences overlapping with one another on a 50 nt segment. Subsequences shorter than 100 nt, i.e., at the 3′ end, were also kept. All subsequences were then folded using the RNA-subopt function in the Vienna RNA package [169] with the setting -e 0. The Vienna RNA package is highly regarded for its ability to accurately predict the secondary structure of non-coding RNA using the minimal thermodynamic energy approach. With this -e 0 setting, multiple folding structures that have the same minimum thermodynamic energy are generated. Using this method with RSmatch, 773 structures were obtained from the eight original *Drosophila roX1* sequences.

RSmatch was used to conduct pairwise comparisons of all 773 RNA structures produced in the process previously described. RSmatch was configured for nucleotide matching scores of 1 and 3 in single-stranded (ss) and double-stranded (ds) regions, respectively. In addition, mismatch scores configuration settings were −1 and 1, in ss and ds regions, respectively. The gap penalty was −6 for both ss and ds regions. This scoring scheme essentially gave more weight on matches in ds regions

TABLE 5.4: *Drosophila roX1* sequences used in this case study

Species	Length	FlyBase Region	Region Coordinates*
D. yakuba	3,433	Cr X	4658396 - 4661828
			3710814 - 3710795
D. simulans	3,439	Cr X	2761962 - 2759151
			2762379 - 2761996
			2759122 - 2758943
			9903425 - 9903446
D. erecta	3,462	Sc 4690	1139892 - 1137083
			1140318 - 1139928
			1137036 - 1136857
D. melanogaster	3,468	Cr X	3755987 - 3754338
			3754043 - 3753143
			3756379 - 3756024
			3754304 - 3754082
			3753108 - 3752929
D. pseudoobscura	3,469	XL group1e	6901185 - 6898994
			6898915 - 6897910
			6897801 - 6897717
			1352025 - 1352045
			10880750 - 10880730
			476212 - 476239
			2910133 - 2910114
D. ananassae	3,493	Sc 13117	695557 - 693154
			693089 - 692300
			692143 - 692065
			692247 - 692215
D. virilis	3,623	Sc 13042	4639617 - 4638455
			4637622 - 4636608
			4638333 - 4637894
			4636532 - 4636064
			4637736 - 4637672
			4637835 - 4637787
			4636035 - 4635995
			4638396 - 4638367
D. mojavensis	3,768	Sc 6328	3900419 - 3899115
			3901467 - 3900566
			3901937 - 3901499
			3902390 - 3902000
			3898736 - 3898624
			3898929 - 3898874
			3898845 - 3898810
			3947396 - 3947375
			246881 - 246900
			700541 - 700522

* Source: www.flybase.org; Cr = chromosome; Sc = scaffold.

TABLE 5.5: Description of 12 *Drosophila* genomes downloaded from
FlyBase

Species	Release #	Release Date	Nucleotides Region	Original Files
D. melanogaster	5.18	5/16/2009	130,430,583	7
D. simulans	1.3	7/24/2008	137,828,247	1
D. erecta	1.3	7/24/2008	152,712,140	1
D. pseudoobscura	2.4	5/19/2009	152,738,921	1
D. yakuba	1.3	7/24/2008	165,693,946	1
D. sechellia	1.3	7/24/2008	166,577,145	1
D. persimilis	1.3	7/24/2008	188,374,079	1
D. mojavensis	1.3	7/24/2008	193,826,310	1
D. grimshawi	1.3	7/24/2008	200,467,819	1
D. virilis	1.2	7/24/2008	206,026,697	1
D. ananassae	1.3	7/24/2008	230,993,012	1
D. willistoni	1.3	7/24/2008	235,516,348	1

than those in ss regions. Three unduplicated FASTA sequences were
identified and extracted from the highest scoring pairwise alignments.

The MXSCARNA [357] package was used to align sequences for the
covariance model (CM) used in the case study. The resulting alignment
was rendered in the Stockholm format with predicted structure annotation. This alignment was input to the cmbuild Infernal package utility
to create a CM.

The CM search utility, cmsearch, was run against a dataset of
Drosophila FASTA sequences. The genomes from 12 *Drosophila* species
(i.e., *D. ananassae, D. erecta, D. grimshawi, D. melanogaster, D. mojavensis, D. persimilis, D. pseudoobscura, D. sechellia, D. simulans, D.
virilis, D. willistoni* and *D. yakuba*) were downloaded from Indiana University's FlyBase [84] (Table 5.5). Many *Drosophila* genomes have not
yet been completely annotated into clearly defined chromosomes. As part
of active research and sequencing efforts, the annotation of *Drosophila*
and other genomes has become richer and more informative.

The Infernal package (version 1.0) utility known as cmsearch was
used to locate structures in *Drosophila* genomes with probability of
matching the constructed CM [270]. To improve computational efficiency, large FASTA sequences were subdivided into smaller, overlapping
subsequences to facilitate independent parallel searching without nega-

tively impacting results. Using a stochastic dynamic programming algorithm, Infernal located and reported secondary structures in *Drosophila* genomes similar to the profile that the CM represented. Given the structural similarity and high score result of the CM search, a *D. sechellia* sequence discovered in the genome scan was predicted to represent *roX1* functional domain characteristics.

5.4.3   Results

The objective of this case study was to classify functional structure elements, i.e., non-coding RNA, in genomes of *Drosophila* species as potential *roX1* homologues. To an extent, portions of an approach previously deployed [201, 202, 203, 238] were used as models. First, eight sequences of *roX1* RNA transcripts were obtained (Table 5.4). Next, a slide-and-fold method to construct RNA structures was executed, as described in Methods. Subsequences of 100 nucleotides (nt) in length or shorter were folded according to their thermodynamic properties using the Vienna RNA package [169]. Adjacent subsequences were overlapped by 50 nt. Non-coding RNA structures can be predicted accurately and efficiently in this way for two reasons:

- Prediction for small ribonucleotide structures is more accurate and efficient than for large ones

- Structures smaller than 50 nt were folded twice as subsequences of two different larger structures, further increasing the probability of obtaining accurate RNA structure predictions.

The Vienna folding package was run with a configuration that yielded multiple RNA structure predictions with the same minimum free energy for a given sequence to further improve folding accuracy. This step resulted in 773 predicted RNA structures.

Species versus species pairwise comparisons were applied using all 773 predicted RNA structures. For computational efficiency, each alignment was run on a separate processor independent of all others using a high performance computation (HPC) cluster, leveraging parallel processing speed-up capabilities [295]. This HPC Sun Microsystems Discovery cluster has 112 AMD Opteron dual-core Linux nodes with 2 GB of RAM per node. The operating system used was Red Hat Enterprise Linux AS release 4, update 8. Approximately 520,000 pairwise alignments were completed in less than five minutes. Each comparison yielded an alignment score. A group of three structures that scored similarly and had similar lengths were selected. RNA structures were obtained from *D. melanogaster*, *D. simulans* and *D. yakuba*.

A covariance model (CM) was created from this group of three structures by first aligning the sequences into the Stockholm format and then running the Infernal cmbuild utility. The complete genomes of eight *Drosophila* species for which the presence of *roX1* ncRNA transcripts had been confirmed were used as targets in CM searches. All complete genomes used were obtained from Indiana University's Fly-Base (http://www.flybase.org) [84]. These genomes were the most current releases at the time the study was conducted (Table 5.5). A CM search located the *roX1* genes precisely where they were known to be present in six *Drosophila* species, i.e., *D. ananassae*, *D. erecta*, *D. melanogaster*, *D. pseudoobscura*, *D. simulans*, and *D. yakuba*. However, the CM search failed to locate the known *roX1* ncRNAs on the remaining two *Drosophila* species, i.e., *D. mojavensis* and *D. virilis*. In five of the six successful searches, the highest scoring result represented a sequence within the known range of the *roX1* genomic coordinates for that species. The sixth successful search on *D. pseudoobscura* produced the third highest scoring search result that represented a sequence within the known range of the *roX1* genomic coordinates for that species. The two highest scores for *D. pseudoobscura* likely represent sequences with conserved *roX1* functionality.

For computationally efficiency, the downloaded genome files were separated into smaller files of 2 million base pairs (Mbp) per file. FASTA sequences larger than 2 Mbp were split into smaller FASTA sequences that overlapped one another by 5,000 base pairs (Kbp) to prevent loss of accuracy. This approach is similar to the RSmatch slide-and-fold approach described in Methods. However, this process was performed with custom Perl scripts. Concurrent cmsearch jobs were run against multiple genome sequences in parallel using a high performance computation (HPC) cluster. A covariance model (CM) search of an entire genome took about 10 minutes.

This case study then focused on classifying potential *roX1* functional structure elements in the genomes of the four fully sequenced *Drosophila* species in which the presence of *roX1* transcripts had not yet been confirmed. These four species were named *D. grimshawi*, *D. persimilis*, *D. sechellia* and *D. willistoni*. The most current release of these complete genomes were obtained from FlyBase [84]. The same CM previously used was used to search for *roX1* functional domains. While scoring results were not significant for three of the four species, a strong score resulted from the search on the *D. sechellia* genome (Table 5.6). This high score shows strong evidence of a *roX1* functional domain in a specific area of the *D. sechellia* genome, namely scaffold_4. Furthermore, in spite of the *D. sechellia*'s incomplete annotation, this result likely indicates that this region of the genome may be located in the X chromosome of *D.*

sechellia. These findings need to be confirmed by wet lab experiments.

To investigate possible *roX1* homology between species, *roX1* gene sequences FBgn0019661 (for *D. melanogaster*) and FBgn0255860 (for *D. sechellia*) were downloaded from FlyBase. A pairwise alignment on the two sequences was performed. Using the DiAlign program [354] with the -n option for nucleic acid sequence comparison, a result of 94% similarity between the two gene sequences was shown. This result indicates high probability of conserved *roX1* functionality between the two species.

In this study a systematic and computationally efficient approach was designed and developed to classify *roX1* RNA structure elements conserved in *Drosophila* species. It consisted of three major steps:

- Comparison of RNA structures among all *roX1* RNAs

- Selection of RNA structure groups significantly associated with those in other species

- Utilization of a highly regarded structure-searching methodology (i.e., covariance models) which, in addition to being highly sensitive and specific, is also very flexible

The stochastic representation of a cluster of RNA structures can be fine-tuned as needed by adding or removing structures from the cluster. Parallel processing contributes to overcoming the burden of lengthy processing times. This method was applied to classifying small RNA structures chiefly because these structures can be classified more accurately than methods that only use thermodynamic minimization. As more powerful RNA structure classification and prediction programs become available, this case study approach can be extended to larger RNA structures.

To compare the effectiveness of different tools, BLAST was compared with Infernal in a search for conserved structural motifs. Since BLAST is not designed to detect covariant base pairs that are critical in an RNA secondary structure, Infernal was expected to perform better. Each of the three sequences from the covariance model were used in FlyBase BLAST to search for homologues in the complete genomes of all 12 *Drosophila* species downloaded from FlyBase. Every homologue detected by BLAST was also detected by Infernal. However, BLAST failed to detect *roX1* evidence in *D. ananassae* and *D. pseudoobscura*, while such evidence was detected by Infernal. This simple experiment provides an insight into the complexity involved in the classification of ncRNA motifs.

By conducting a homology scan on a complete genome or species chromosome, a researcher can confirm whether a functional domain is present throughout that genome or species chromosome. A stem-loop

TABLE 5.6: Summary of homologues found in seven *Drosophila* species

Genome Scanned	CM Score	FlyBase Region	Region Coordinates	Strand	In roX1?
D. melanogaster	88.84	Chromosome X	3753295 - 3753232	−	Yes
D. yakuba	88.69	Chromosome X	4661475 - 4661538	+	Yes
D. simulans	88.1	Chromosome X	2759303 - 2759240	−	Yes
D. sechellia	88.1	Scaffold 4	2954091 - 2954154	+	N/A
D. erecta	72.78	Scaffold 4690	1137235 - 1137172	−	Yes
D. ananassae	32.26	Scaffold 13117	692432 - 692373	−	Yes
D. pseudoobscura	29.4	Unknown group 410	14965 - 14898	−	No
D. pseudoobscura	28.28	XL group1e	6898105 - 6898042	−	Yes
D. pseudoobscura	29.11	Unknown group 260	63165 - 63089	−	No

N/A = Not available

structure was previously predicted in *roX1* RNA on the X chromosome of *D. melanogaster* [352] and determined that this structure was conserved in several species of *Drosophila* [288]. This study confirms that among seven different *Drosophila* species, the *roX1* functional domain is only present on the X chromosome and is absent from all chromosomes other than X. The maturation of genome annotation and the translations of scaffold regions into chromosome regions will reveal whether this observation continues to hold.

5.4.4 Conclusion

RSmatch and Infernal were demonstrated to be effective tools in discovering patterns of novel ncRNAs. Homology searching is common in bioinformatics, yet some of the most popular homology search methods such as BLAST and FASTA are often the least accurate [122]. For non-coding RNA, homology searching is more challenging than a sequence homology search. This is due to intramolecular covariant base pairs in ncRNA that are conserved to a higher degree with respect to their primary structures, i.e., their nucleotide sequence.

An Infernal search requires considerable computer run time [270]. Freyhult, et al., estimated that with a query for the transfer RNA (tRNA) type of ncRNA, Infernal would take about 96 days to search the entire human genome on a single processor [122]. Innovative methodologies including HMM filtering and sequence-based heuristics [389, 409] have been employed to improve computational efficiency. In our study, parallel processing with a high performance computing cluster was used for speed-up and improved throughput.

Whole genomes of all 12 sequenced species of *Drosophila* were scanned. All 12 are believed to have a common ancestor that existed about 40 million years ago [288]. Phylogenetic relationships are based on the premise that species that evolved relatively recently will have more genetic similarities than species that evolved earlier. As a result of this study, the presence of the *roX1* ncRNA was verified as previously reported by other authors in six *Drosophila* species. In addition, strong evidence of the presence of *roX1* in *D. sechellia* was found, which was not known to be previously reported.

5.5 CASE STUDY IN BIOLOGICAL DATA MINING

As an example of biological data mining, a case study in the mining of a three-dimensional RNA motif is presented. Artificial intelligence tools are used to find motifs in DNA, RNA and proteins. In the study, a computational tool for finding RNA tertiary motifs in genomic sequences was designed and developed. Specifically, this tool predicted genomic coordinate locations for coaxial helical stackings in three-way RNA junctions. These predictions were provided by CSminer, a tertiary motif search package that utilized two versatile methodologies: random forests and covariance models. A coaxial helical stacking tertiary motif occurs in a three-way RNA junction where two separate helical elements are aligned on a common axis to form a pseudocontiguous helix that provides thermodynamic stability to the RNA molecule. The CSminer tool used a genome-wide search method based on covariance models to find genomic regions that may contain a coaxial helical stacking tertiary motif. CSminer also used a random forests classifier to predict whether the genomic region contained the tertiary motif. Experimental results demonstrated the effectiveness of the CSminer approach.

5.5.1 Background

It is important for bioinformatics researchers to develop pattern discovery tools that leverage increasingly powerful computational methodologies in critical life science research. In this case study, CSminer (i.e., Coaxial helical Stacking miner), predicted locations or genomic coordinates of coaxial helical stackings in genomes. A coaxial helical stacking occurs in an RNA tertiary structure where two separate helical elements are aligned on a common axis and form a pseudocontiguous helix [215] at an RNA junction. An RNA junction is an important non-coding RNA (ncRNA) loop structure that forms where three or more helices meet. Coaxial helical stacking tertiary motifs may occur in several large RNA structures, including group II introns [366], large ribosomal subunits [30, 324, 398], pseudoknots [2], and transfer RNA (tRNA) [206]. Coaxial helical stackings provide thermodynamic stability to the RNA molecule [205, 375], and reduce the separation between loop regions within junctions [3]. Coaxial helical stacking is also involved in long-range interactions in many RNAs [405] and are essential features in a variety of other RNA junction topologies.

Our focus was on the three-way RNA junction, although many RNA junctions exist in four-way and higher forms. The topologies of known three-way RNA junctions have been studied extensively [230]. Three-way RNA junctions that contain coaxial stacking are classified into three topological families called A, B and C, depending on the orientation of the helix that is not involved in the coaxial stacking and on the lengths of the unpaired base regions separating the helices. Figure 5.1 illustrates these three topological families.

Each three-way RNA junction in Figure 5.1, has three helices labeled P1, P2 and P3. In a helix region, bases are grouped in standard Watson-

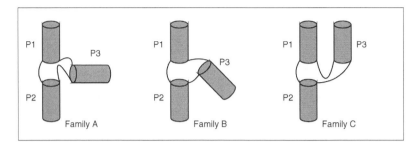

FIGURE 5.1: Three RNA topological families, A, B, and C, of three-way RNA junctions containing coaxial helical stacking.

Crick pairings. In the examples in the figure, P1 and P2 are presumed to be coaxially stacked. There is no presumption in any of these examples about the positions of the 5′ end and the 3′ end of the RNA molecule. In bioinformatics literature, the helix in a three-way RNA junction nearest to the 5′ and 3′ ends is generally considered to be the first helix. The number of a helix does not necessarily indicate its position relative to the 5′ and 3′ ends of the RNA molecule. In each three-way RNA junction in Figure 5.1, the unpaired base region between P1 and P2 is called loop strand J12, the unpaired base region between P2 and P3 is called loop strand J23, and the unpaired base region between P3 and P1 is called loop strand J31.

The following are characteristics of each of the three topology families of three-way RNA junction:

In RNA topology family A:

- Loop strand J12 is the shortest of the three inter-helical loop strands.

- Loop strand J31 is typically shorter than loop strand J23.

- P3 is roughly perpendicular to the coaxial stacking of P1 and P2.

In RNA topology family B:

- The three loop strands J12, J23 and J31, are all approximately the same length.

- P3 is oriented closer to P2 than to P1.

In RNA topology family C:

- Loop strand J12 is the shortest of the three inter-helical loop strands.

- Loop strand J31 is typically longer than loop strand J23.

- P3 is oriented closer to P1 than to P2.

It is believed that the function of RNA is closely associated with its 3D structure, which, by virtue of canonical Watson-Crick base pairings (i.e., AU, GC) and wobble base pairing (i.e., GU), is largely determined by its secondary structure [227, 301, 308]. Several tools are available for secondary structure prediction and ncRNA search. One of the most highly regarded is Infernal [270], discussed in this chapter's previous case study. A wide variety of statistical analysis approaches, in particular, ensemble-based approaches, have been successful in life science applications. Laing, et al., applied an ensemble-based approach, specifically random forests, to predict coaxial helical stacking in RNA junctions [215]. This random forests classifier is called Junction Explorer.

In this case study, the functionality of Infernal was extended to create the CSminer tool to predict a tertiary RNA motif, i.e., a coaxial helical stacking, in a genome. This was accomplished by invoking Junction Explorer [215] within Infernal to evaluate each significantly high-scoring Infernal result and report the coaxial stacking status of these results. The secondary structure in each Infernal result is formatted into Connectivity Table (CT) format and evaluated by Junction Explorer to confirm the pattern match.

5.5.2 Methods

Laing, et al., studied 110 distinct three-way RNA junctions that were confirmed in crystal structures [215]. Each of these 110 unique junctions was verified in one of 32 crystal structure models in the Protein Data Bank (PDB) [38]. The majority or 75% of these 110 three-way RNA junctions were found in the complex ribosome subunit molecules (rRNA), i.e., 51% in 23S rRNA, 20% in 16S rRNA and 4% in 5S rRNA. Kingdoms represented in these 32 PDB samples were bacteria, archaea, animalia and plantae.

There was no dominant topological configuration among these 110 three-way RNA junctions in that 47% were categorized as family C, 35% as family A and the remaining 18% as family B [215]. For each of these 110 junctions, the coaxial helical stacking status was known. The coaxial stacking status of each junction was described as one of four possibilities: H_1H_2, H_1H_3, H_2H_3 or none, where H_xH_y indicated that helix H_x shared a common axis with helix H_y. The helix identified as H_1 was the first helix in the three-way RNA junction, as described below.

A three-way RNA junction is described by three subsequences [215]. For each subsequence, base coordinates and base values (i.e., A, C, G,

U) are known. The starting and ending coordinates of each subsequence indicate the 5' and 3' ends of the subsequence, respectively. Unpaired bases of each subsequence are referred to as parts of the loop regions of the junction and are used to help determine the coaxial helical stacking status of the junction as described later. The three-way RNA junction formed by these subsequences includes unpaired bases of the loop region, terminal base pairs of the three helices and the second to last base pairs of the three helices, as follows. The 5' end of the first subsequence is the 5' base of the second to last base pair of helix H_1. The 3' end of the first subsequence is the 5' base of the second to last base pair of helix H_2. Similarly, the 5' end of the second subsequence is the 3' base of the second to last base pair of helix H_2, and the 3' end of the second subsequence is the 5' base of the second to last base pair of helix H_3. It follows that the 5' end of the third subsequence is the 3' base of the second to last base pair of helix H_3, and the 3' end of the third subsequence is the 3' base of the second to last base pair of helix H_1. The length of each subsequence is at least 4. The first two bases of each subsequence are part of one helix and the last two bases of that subsequence are part of the next sequential helix. There may be no unpaired bases between two helices that share a subsequence. In this case, the subsequence length is minimal, i.e., 4.

Figures 5.2 and 5.3 illustrate a three-way RNA junction in nucleotide positions 5 through 49 of chain A in PDB molecule 3E5C, i.e., "Crystal Structure of the SMK box (SAM-III) Riboswitch with SAM." This junction is known to have a coaxial helical stacking identified as H_2H_3, i.e., helices H_2 and H_3 share a common axis.

The secondary structure plot for the RNA sequence is shown in Figure 5.2 produced using VARNA [89]. In this figure, the junction is enclosed within a dashed line. The first subsequence starts at position 5, ends at position 10 and consists of the bases CCGAAA. The second subsequence of the three-way RNA junction starts at position 34, ends at position 41 and consists of the bases UUGUAACC. Finally, the third subsequence of the junction starts at position 46, ends at position 49 and consists of the bases GGGG. Unpaired bases in the loop region are not part of the terminal base pairs of the three helices. In this figure, helices H_2 and H_3 are shown to be coaxially stacked with the aid of a super-imposed bar.

Figure 5.3 was obtained using Jmol [165] and presents a three-dimensional representation of the same RNA molecule shown in Figure 5.2, i.e., positions 5 through 49 of chain A in PDB molecule 3E5C. This figure represents the crystal structure 3D coordinates of the 976 atoms that comprise this RNA molecule. Helix H_1 base positions 5 and 6 are identified, as are helix H_2 positions 34 and 35, and helix H_3 positions

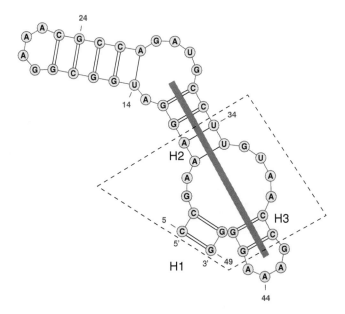

FIGURE 5.2: Secondary structure plot produced by VARNA of nucleotides 5 through 49 of chain A from PDB molecule 3E5C. The three-way RNA junction is enclosed by a dashed line. The three helices of junction are labeled H_1, H_2 and H_3, respectively.

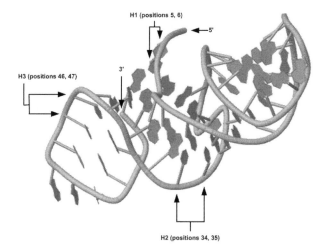

FIGURE 5.3: Three-dimensional plot produced by Jmol of nucleotides 5 through 49 of chain A from PDB molecule 3E5C. Two positions from helix H_1, helix H_2 and helix H_3 are identified. Helices H_2 and H_3 are seen to be coaxially stacked, i.e., sharing a common axis.

46 and 47. The coaxial helical stacking of H_2 and H_3 is apparent. In addition, Jmol provides interactive viewing of 3D figure rotations. By rotating and viewing the figure from any angle, the coaxial helical stacking becomes more visible.

A coaxial helical stacking motif in a three-way RNA junction can be predicted by Junction Explorer which has been trained using certain specifically chosen features readily available in the secondary structures of known three-way RNA junctions, i.e., the 110 element dataset described above. Collecting appropriate features for motif prediction is among the difficult yet important challenges in bioinformatics, pattern recognition and machine learning. Features used for this study were previously collected [215]. Figure 5.4 shows a hypothetical three-way RNA junction that illustrates features used in the random forests classifier. Helix regions, consisting of paired nucleotides, are identified as H_1, H_2 and H_3. Loop regions consisting of unpaired nucleotides are identified as J12, J23 and J31.

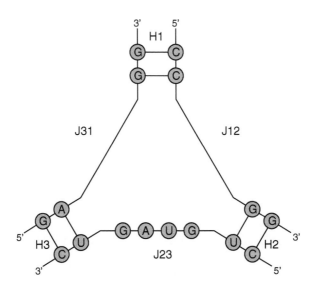

FIGURE 5.4: Hypothetical three-way RNA junction to illustrate features used by random forests classifier.

Table 5.7 describes 15 features used to train Junction Explorer employed in the study. Feature values were derived from attributes of known three-way RNA junctions. A hypothetical three-way RNA junction is shown in Figure 5.4. Features used are based on three principles.

- A short loop region in the three-way RNA junction, i.e., the unpaired strand between adjacent helices, is more likely to be associated with a coaxial helical stacking. For this reason, the sizes or lengths of the three loop regions (i.e., the numbers of unpaired nucleotides in the three loop regions) of a three-way RNA junction are used as features as well as the manner in which these three sizes relate to one another, e.g., the minimum, median and maximum of the three sizes.

- Consecutive unpaired adenine bases tend to interact via hydrogen bonding with the minor groove of a neighboring helix. This common interaction, known as A-minor motif, stabilizes contacts between RNA helices. In fact, the A-minor motif is the most common tertiary interaction in the large ribosomal subunits. For this reason, information about consecutive unpaired adenine bases is treated as a feature.

- Thermodynamic free energy associated with the base pairs at the helix termini and the loop regions between adjacent helices is used as a feature. As thermodynamic free energy declines in a conformation, stability increases.

In total, 15 features were used for coaxial helical stacking prediction to train Junction Explorer used in this case study.

Thermodynamic free energy of two adjacent helices was determined by the length or size of the unpaired nucleotide loop region between the two helices, denoted $LoopSize$, as follows:

- When $LoopSize$ was 0, the free energy values were taken from the table of RNAstructure [309].

- When $LoopSize$ was 1, the free energy values were taken from the table of RNAstructure plus 2.1. If two possible values existed, the smaller one was taken.

- When $LoopSize$ ranged from 2 to 6, the free energy values were calculated as follows: $a + b*LoopSize + c*h$ where $a = 9.3$, $b = -0.3$, $c = -0.9$, $h = 2$.

- When $LoopSize$ was greater than 6, the free energy values were calculated as follows: $a + 6*b + 1.1*ln(LoopSize/6) + c*h$ where $a = 9.3$, $b = -0.3$, $c = -0.9$, $h = 2$.

TABLE 5.7: Features used in random forests classifier

Feature Name	Feature Value	Feature Description
A(J12)	0	Adenine bases in loop region J12
A(J23)	1	Adenine bases in loop region J23
A(J31)	0	Adenine bases in loop region J31
$\Delta G(H_1, H_2)$	-1.4	Thermodynamic free energy of helices H_1 and H_2
$\Delta G(H_2, H_3)$	6.3	Thermodynamic free energy of helices H_2 and H_3
$\Delta G(H_1, H_3)$	-2.1	Thermodynamic free energy of helices H_1 and H_3
|J12|	0	Length of J12 loop region in bases
|J23|	4	Length of J23 loop region in bases
|J31|	0	Length of J31 loop region in bases
Min(|J12|,|J23|,|J31|)	0	Minimum of 3 loop region lengths
Med(|J12|,|J23|,|J31|)	0	Median of 3 loop region lengths
Max(|J12|,|J23|,|J31|)	4	Maximum of 3 loop region lengths
Min(|J12|,|J31|)	0	Minimum of J12 and J31 loop region lengths
Min(|J12|,|J23|)	0	Minimum of J12 and J23 loop region lengths
Min(|J23|,|J31|)	0	Minimum of J23 and J31 loop region lengths

The CSminer program combines Junction Explorer described above with Infernal [270], discussed in this chapter's previous case study. Complete nucleotide sequences were extracted, starting at the 5′ end and ending at the 3′ end, from the Protein Data Bank (PDB) [38] for all 110 known three-way RNA junctions described above. Using RNAview [408] for guidance, the 110 secondary structures were manually evaluated and 31 were selected based on similarity of length and general secondary structure. When searching a genome for ncRNA secondary structure motif matches, Infernal uses a covariance model (CM) comprised of several similar ncRNA secondary structures. As Infernal builds a CM, it takes into account the differences among the structures used in building the model, and shapes the model with statistical representations of these differences. Since the purpose of Infernal is to find structures similar to the CM, the structures that comprise the CM must also be similar. The more members that make up the CM, the more effective the model becomes. Even though a CM can be built using a single ncRNA structure, such a model would have significantly reduced effectiveness in locating similar structures.

The 31 selected secondary structures were clustered using RNAforester [167]. RNAforester clusters structures based on secondary structure similarity. Six three-way RNA junctions with similar secondary structures were grouped into a high-scoring cluster by RNAforester. These six junctions had similar secondary structures and known coaxial helical stackings. The six junctions belonged to PDB molecules with identifiers 2GDI, 2CKY, 2AVY, 1S72, 2AW4 and 2J01, respectively, and formed the CM used in the case study.

A Stockholm format multiple sequence alignment is required to create an Infernal CM. The structure alignment provided by RNAforester was manually extracted, along with the consensus secondary structure. These established the required Stockholm format multiple sequence alignment (Figure 5.5). The Stockholm format is a multiple alignment of ncRNA sequences together with the consensus secondary structures of the aligned sequences. The secondary structures are shown in dot-parentheses notation, in which dots represent bases and parentheses represent base pairs. The CM was created from the constructed Stockholm format using Infernal's CMbuild utility [270].

Infernal's CMsearch utility was extended to execute a trained random forests classifier, i.e., Junction Explorer, whenever an ncRNA secondary structure similar to the covariance model was detected during genome-wide searches. The resulting program was named CSminer.

The trained random forests classifier was capable of predicting a coaxial helical stacking in a three-way RNA junction within the ncRNA secondary structure detected by CMsearch. Breiman designed the ran-

```
# STOCKHOLM 1.0
#=GF ID 2GDI(94) (riboswitch)     length=74
#=GF ID 2CKY(96) A. thaliana      length=73
#=GF ID 2AVY(82) E. coli          length=71
#=GF ID 1S72(21) H. marismortui   length=73
#=GF ID 2AW4(22) E. coli          length=71
#=GF ID 2J01(23) T. thermophilus  length=71
#=GF ID COAX_Model_19 (CM model name)
#=GF alignment and consensus structure by RNAforester software

2GDI_94.ct          CUCGGGGU----GCC-CUUCUGCGUGAAGGCUGAGAAAUACCCGUAUCACCU-GA
2CKY_96.ct          ACCAGGGG----UGC--UUGUUCAC-AGGCUGAGAAAGUCCCU-UUGAACCU-GA
2AVY_82.ct          UUAUCCUUUGUUGCCAGCGGUCCGGCCGGCGGGAACUCA-A-AGGA--G--ACUG-C-
1S72_21.ct          GACAAGAUUGAAGCG--UGCCGAAAG-GCACGUGG-A-AGUCUG--UU-AGAGUU-
2AW4_22.ct          GGCAGGUUGAAGGU--UGGGUAACA-CUAACUGG-A-GGACCG--A--ACCGAC-
2J01_23.ct          GCCAGGGUGAAGCU--GGGGUGAGA-CCCAGUGG-A-GGCCCG--A--ACCGGU-
#=GC SS_cons        ((.((((((((((((..(((.....)))))))).).)).)))).....((((.((

2GDI_94.ct          UC-UGGAUAAUGCCAGCGUAGGG-AA--G
2CKY_96.ct          AC-AGGGUAAUGCCUGCGCAGGG-AGUGU
2AVY_82.ct          CA-GUGAU--AA-ACUGGA-GGAAGGUGG
1S72_21.ct          GGUGUCCUACAAUACCCUC-UCG-UGAUC
2AW4_22.ct          UA-AUGUUGAAAAAUUAGC-GGA-UGACU
2J01_23.ct          GG-GGGAUGCAAACCCCUC-GGA-UGAGC
#=GC SS_cons        .(.(((......))))).).))).)...))
//
```

FIGURE 5.5: Stockholm format multiple sequence alignment of ncRNA molecules from six samples recorded in the PDB with identifiers 2GDI, 2CKY, 2AVY, 1S72, 2AW4 and 2J01.

dom forests classifier to be comprised of numerous classification and regression trees (CARTs) [51], each of which is formed by a small random subset of 4 (i.e., the square root) of the 15 features. Each CART is capable of contributing a "better than random opinion" about the coaxial helical stacking prediction of an unknown or unlabeled input. By consolidating all opinions from all CARTs, i.e., by tallying all "votes," the random forests classifier is able to predict the coaxial helical stacking status of the three-way RNA junction.

It takes constant time for the random forests classifier to make predictions, and the space used by the random forests classifier is independent of the genome length. Thus, the space and time complexities of CSminer are the same as Infernal. Specifically, the space complexity of CSminer is $O(L^2M)$ and the time complexity is $O(L^3M)$, where L is the genome length and M is the number of states in the stochastic context-free grammar represented by the covariance model (CM) [104]. Run times varied from 1 second to 1 hour 21 minutes and 48 seconds depending on the size of the target genome and the length of the aligned structures in the CM.

5.5.3 Results

A series of experiments were conducted to evaluate the pattern discovery effectiveness of the approach presented in this case study. In the first experiment, CSminer was run against the complete genome of *Deinococcus radiodurans*, i.e., GenBank ID NC_001263.1, obtained from the NCBI

```
CM: COAX_Model_19
>gi|15805042|ref|NC_001263.1|

  Plus strand results:

  Query = 1 - 74, Target = 251047 - 251117
  Score = 48.18, GC = 62

  Coax status = H2H3

            ((,<<<<<-<<<<<<-<<_____>>>>>>>->->->>>>,,,<<--<-<<<<_____>
       1 gcCaGGguGGaGgcCuggGUacgaccGgcUgGCAagcCCgauACCGacuggugaUAAAAc 60
            :CCAGG::G+A :CC ::GU++ A::GG: GG A:G:CCGA ACCG :+ :UG U+AAAC
       251047 ACCAGGUUGAAACCC-CCGUGACAGGGGGCGG-AGGACCGA-ACCGGUGCCUGCUGAAAC 251103

       >>>->->>,,,,))
          61 acccgcGGguGagc 74
             A: C:CGG+UGAG:
          251104 AGUCUCGGAUGAGU 251117
  //
```

FIGURE 5.6: CSminer's prediction result on the genome of *D. radiodurans*. The detected ncRNA sequence contains nucleotides from positions 251047 through 251117 (with a length of 71 nt) on the plus strand of the genome. Parentheses and brackets above the nucleotides represent base pairs constituting the secondary structure of the ncRNA sequence. The coaxial helical stacking status of H_2H_3 indicates that the ncRNA secondary structure has a three-way RNA junction, and that this junction was predicted to contain a coaxial helical stacking motif of type H_2H_3.

GenBank database [37]. An ncRNA tertiary motif, i.e., H_2H_3 coaxial helical stacking, was detected between positions 251047 and 251117 on the plus strand of the genome (Figure 5.6). This three-way RNA junction was predicted by CSminer to contain a coaxial helical stacking. The coaxial helical stacking was of type H_2H_3 (i.e., helix H_2 and helix H_3 aligned with a common axis).

This CSminer prediction result is validated as follows. Based on NCBI BLAST [13] and manual analysis, *D. radiodurans* is related to PDB molecule 1NKW. Specifically, the chain 0 nucleotide sequence was downloaded from the PDB for the 1NKW structure. Using NCBI BLAST, this downloaded sequence was located in the whole genome of *D. radiodurans* from positions 251047 through 251117 on the plus strand. These positions are consistent with those shown by CSminer where the motif was detected (Figure 5.6). Furthermore, based on previous analysis [215], this region of the 1NKW structure contains a three-way RNA junction with a coaxial helical stacking of type H_2H_3, which is what CSminer reports.

Table 5.8 presents the successful search results from 12 CSminer experiments using three covariance models (CMs) where the CMs were built using the techniques described in the previous subsection. A search is considered successful when the Infernal CMsearch score is higher than 30. The table contains the following columns:

- "Model PDB Identifications" shows the PDB molecules from which RNA sequences of known coaxial helical stacking ncRNA tertiary motifs were extracted, aligned and used to build a CM for the CSminer genome search. The source species from which the PDB molecules come can be found in Table 5.9.

- "Genome Identification" shows the accession number representing the genome sequence searched by CSminer with the respective CM formed with RNA sequences extracted from the PDB identifications shown in the first column. Note that the genome searched with a CM is different from the genomes or species from which RNA sequences were extracted and used to build the CM.

- "Species" shows the name of the species corresponding to the genome identification in the second column.

- "Positions" and "Strand" columns show search result positions in the genome sequence where CSminer predicts the location of a coaxial helical stacking and DNA strand, positive or negative, to which the search result positions pertain. Note that positions increase from low to high for a positive strand search result, and decrease from high to low for a negative strand search result.

- "Status" shows the motif type predicted (H_1H_2, H_1H_3 or H_2H_3), where H_xH_y indicates that helix H_x shares a common axis with helix H_y.

- "Validated" shows whether the predicted result described by the fourth and fifth columns is validated by known crystal structure evidence in the PDB database. Where there is no available crystal structure evidence (i.e., the column shows "No"), the predicted result needs to be validated by wet lab experiments.

CSminer was applied to the complete genome of *Thermus thermophilus*, i.e., GenBank ID CP002777.1. An ncRNA tertiary motif, i.e., H_1H_2 coaxial helical stacking, was detected between positions 14310 and 14384 on the plus strand of the genome. This result was validated by cross-checking the result coordinates against the known motifs in PDB molecule 2J01. This result is listed as "validated" in Table 5.8.

CSminer was also applied to the complete genome of *Haloarcula marismortui*, i.e., GenBank ID NC_006397.1. Two ncRNA tertiary motifs, i.e., H_1H_2 and H_2H_3 coaxial helical stackings, were detected between positions 2771 and 2656 on the minus strand of the genome. This result was validated by cross-checking the result coordinates against the known motifs in PDB molecule 1S72. This result is also listed as "validated" in Table 5.8.

TABLE 5.8: Successful search results from 12 CSminer experiments

Model PDB Identifications	Genome Identification	Species	Positions	Strand	Status	Validated?
1NKW,1S72,2AW4	CP002777.1	T. thermophilus	14310-14384	+	H_1H_2	Yes
	NC_013209.1	A. pasteurianus	1536843-1536769	−	H_1H_2	No
	NC_009484.1	A. cryptum	2585998-2585924	−	H_1H_2	No
	NC_016582.1	S. bingchenggensis	9707198-9707272	+	H_1H_2	No
1S72,2AVY,2AW4,	NC_001263.1	D. radiodurans	251047-251117	+	H_2H_3	Yes
2CKY,2GDI,2J01	NC_013209.1	A. pasteurianus	1731899-1731829	−	H_2H_3	No
	NC_009484.1	A. cryptum	2009814-2009744	−	H_2H_3	No
	NC_016582.1	S. bingchenggensis	7289052-7289122	+	H_2H_3	No
1NKW,2AW4,2J01	NC_006397.1	H. marismortui	2771-2656	−	H_1H_2 H_2H_3	Yes
	NC_013209.1	A. pasteurianus	1538830-1538717	−	H_1H_2 H_2H_3	No
	NC_009484.1	A. cryptum	2587983-2587870	−	H_1H_2 H_2H_3	No
	NC_016582.1	S. bingchenggensis	9704954-9705067	+	H_1H_2 H_2H_3	No

In addition, experiments were conducted by selecting three bacterial genomes that are closely related phylogenetically to species represented in the covariance models (CMs) used in this study. The three bacterial genomes selected were *Acetobacter pasteurianus* (GenBank ID NC_013209.1), *Acidiphilium cryptum* (GenBank ID NC_009484.1) and *Streptomyces bingchenggensis* (GenBank ID NC_016582.1). Experiments were also conducted to see whether these CMs would produce any meaningful search results in other biological kingdoms. We selected two viral genomes, one animal genome, one fungus genome and one protista genome. The two viral genomes selected were human immunodeficiency virus 1 (GenBank ID NC_001802.1) and human immunodeficiency virus 2 (GenBank ID NC_001722.1). The animal genome selected was *Drosophila melanogaster* chromosome X (GenBank ID NC_004354.3). The protista species chosen, *Plasmodium falciparum* (GenBank ID NC_004317), is a protozoan parasite and one species of *Plasmodium* that causes malaria in humans. The fungus species chosen, *Saccharomyces cerevisiae* (GenBank ID NC_001136), is one of the most intensively studied eukaryotic model organisms in molecular and cell biology, much like *Escherichia coli*.

This ncRNA tertiary motif prediction method was performed on these additional genomes using CSminer. For each of the additional bacterial genomes, a motif was predicted. None of the additional bacteria organisms selected is represented in the PDB. Therefore, it cannot be confirmed that these predicted results are in fact coaxial helical stackings. These predictions are left to be validated with wet lab experiments. All the predicted results are listed in Table 5.8.

For the two viral genomes, the animal genome, the fungus genome and the protista genome, no motif was predicted. This was likely an indication that the motifs of interest were specific to the genomes or species represented in the covariance models (CMs) used by CSminer (cf. Table 5.9).

Two segments from PDB molecules 2GDI and 2CKY (used in the third CM in Table 5.8) are members of Rfam family RF00059, i.e., the "TPP riboswitch, also known as the THI element and Thi-box riboswitch" [142]. However, none of the validated or non-validated results was found to overlap with a known RNA gene member of some Rfam family. This is not unusual. When CSminer finds a match to the CM used in its search, that match need only be similar to a small number of structures comprising the CM. The match need not be similar to every structure of the CM.

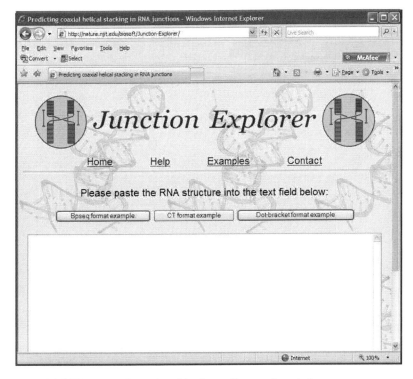

FIGURE 5.7: Junction Explorer Screenshot 1: home page.

5.5.4 Junction Explorer

The random forests classifier, i.e., Junction Explorer used by CSminer is made available as a Web server [215] which helps reveal the potential presence of coaxial helical stackings in an RNA molecule given its secondary structure. The screenshot in Figure 5.7 illustrates the home page of the Junction Explorer Web server.

Junction Explorer accepts as input an RNA primary sequence along with its secondary structure in one of three formats: Bpseq, CT and Vienna dot-bracket (see screenshot in Figure 5.8). The user follows three steps when running the Junction Explorer server:

- Paste an RNA sequence and its secondary structure represented in one of three formats into the blank text field of the server (or simply click the "Examples" button above the text field to retrieve an example RNA molecule).

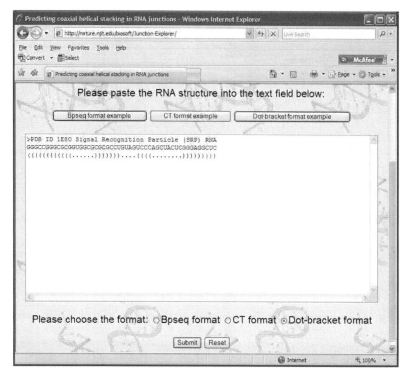

FIGURE 5.8: Junction Explorer Screenshot 2: sample input in dot-bracket format.

- Select the appropriate format option that corresponds to the RNA sequence and structure provided.

- Click the "Submit" button.

After accepting the RNA molecule, Junction Explorer identifies and locates the junctions in the molecule and predicts the presence of coaxial helical stackings. The topology (family) of each junction in the input structure is predicted. Junction Explorer presents a detailed report of the junction type, location, loop regions and the predicted coaxial helical stacking of each identified junction in the molecule (see Figure 5.9).

A graphical display of the predicted topology (family) for each junction is presented (see Figure 5.10). The viewer can easily visualize the stack and topology (family) configuration in each junction.

In most cases, Junction Explorer displays the output on the Web browser promptly. When the size of the input data is large, processing may be time consuming. In that case, a hyperlink identifying where

TABLE 5.9: Species and kingdom for each PDB molecule used to build covariance models employed by CSminer

PDB ID	Species	Kingdom
2CKY	*Arabidopsis thaliana*	Plantae
1NKW	*Deinococcus radiodurans*	Bacteria
2AVY	*Escherichia coli*	Bacteria
2AW4	*Escherichia coli*	Bacteria
2GDI	*Escherichia coli*	Bacteria
1S72	*Haloarcula marismortui*	Archaea
2J01	*Thermus thermophilus*	Bacteria

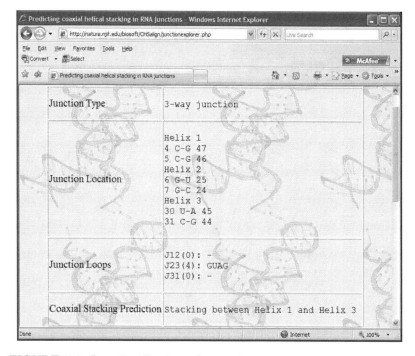

FIGURE 5.9: Junction Explorer Screenshot 3: details of junction type, junction location, junction loop regions and coaxial helical stacking prediction.

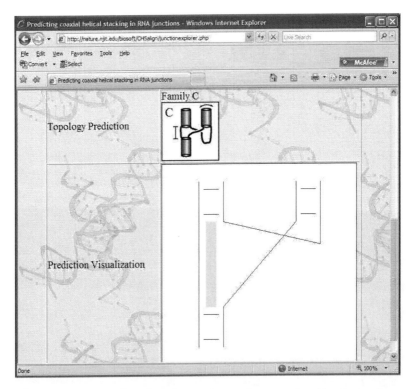

FIGURE 5.10: Junction Explorer Screenshot 4: visualizations of predicted topology (family).

the results will be available is provided to the user. If a pseudoknot is detected in the input structure, Junction Explorer utilizes the K2N tool [337] for removal prior to performing the junction identification and prediction.

5.5.5 Conclusion

This case study demonstrated that CSminer, by combining the strengths of the genome-wide ncRNA Infernal search tool with an ensemble-based random forests classifier, i.e., Junction Explorer, was an effective biological data mining instrument. Among the growing number of ensemble-based methodologies, the random forests method is among the most accurate. This method adds significant functionality to Infernal. Effective mining of coaxial helical stacking motifs in genomes will help to further unravel the mysteries of non-coding RNA. Much remains unknown in this exciting research area. Our study conclusion was that genome-wide mining of coaxial helical stacking RNA motifs is feasible and cost effective.

CHAPTER 6

Biological Network Inference

CONTENTS

6.1   INTRODUCTION

Increasing interest in biological network inference stems from the emergence of a new era of high-throughput measurement and systems biology subsequent to the human genome project completion in 2000 [159, 217]. Systems biology consists of the computational and mathematical modeling of complex biological systems. It is an interdisciplinary field of study that uses a more holistic approach as compared to the more traditional reductionist techniques of biological research. A much different engineering approach to biological research can be followed now because inference of biological networks by reverse engineering has the potential to deliver a far greater understanding of various complex biological systems.

Microarray technology and other high-throughput, next generation sequencing methods measure values of gene expression under different conditions at regular time intervals. The collection of measured values is then gathered into datasets for analysis by genomics researchers for network inference analysis.

Many types of biological networks exist, including metabolic, protein-protein interacting (PPI), signaling and transcriptional. Few such networks are thoroughly understood, even for the simplest bacteria. Systems biology, in this sense, is still in its infancy [267].

A metabolic network is the set of metabolic and physical processes that determine the properties of a cell. A metabolic process consists of a biochemical transformation among cellular molecules (metabolites) that turns one molecule (substrate) into another molecule (product) under the action of an enzyme. Metabolic interactions can be inferred based on product molecular weights [52]. It is also possible to infer correlations between pairs of metabolites by studying multiple samples, for instance in a time series fashion. To provide helpful visualization of inferred networks, biology-specific graphical tools are being developed [187].

In the study of proteomics, a protein is seen interacting with another protein to form a modified version of either of the two protein products. The indications that some of these modified products appear to be cancer-related [327] generate much research interest. Using various reverse engineering approaches with proteomics databases, protein-protein interaction (PPI) networks can be computationally inferred.

Networks describing intercellular signaling systems are inferred by evaluating high-throughput cellular response profiles [265]. After applying multiple perturbations such as drug combinations, quantitative cell biology models are constructed from resultant transcript and protein product information. Non-linear ordinary differential equations (ODEs) are supplemented with heuristics to improve inference efficiency. These inferred network models provide valuable insight into the mechanics of biological signaling systems.

Life sustaining functions in cells depend on tightly regulated expressions of genes. Studies show that one gene's expression as messenger RNA (mRNA) may determine the subsequent expression of another gene. In that case, one gene is said to regulate the other gene's expression and the regulator gene is called a transcription factor (TF) gene. When a gene regulates another gene, the first gene is known as a regulator gene and the second is known as a target gene. The regulator gene's mRNA transcript, when translated into protein, is a transcription factor (TF) of a target gene since it plays a role in the transcription or expression of the target gene.

A gene regulatory network (GRN) describes interactions and expression dependencies among genes in a cell [46]. Gene expression is measured by sequencing RNA transcripts in a cell. Collectively, all transcripts present in a cell at one time are known as a transcriptome. The complete sequencing of the transcriptome is called the gene expression profile. Various stages in the development of the gene expression profile are gathered in the gene expression omnibus database (GEO) [33, 107]. Genetic biologists studying gene expression will disturb or perturb the steady state in a cell by applying a physical or chemical change, thereby invoking a response in the form of increased gene expression. Following a controlled perturbation, and at specific time intervals, gene expression profiles are recorded in a time series analysis dataset for evaluation. Gene expression profile datasets of the cell in a steady state are also created for comparative analyses.

In this chapter, various aspects of inferring gene regulatory networks (GRNs) are presented.

6.2 GENE REGULATORY NETWORK INFERENCE

There is intense interest in applying computational power to help explore gene regulatory causation, that is, one gene causing the expression activity of another gene. It is believed that gene expression is closely controlled by a regulatory mechanism that increases or decreases gene expression depending on ever-changing conditions. Studies show that a regulator gene's expression may result in the subsequent increased or reduced expression of a target gene. A regulator gene is said to up-regulate or down-regulate a target gene's expression.

Every living organism orchestrates a coordinated response to external stimuli such as starvation, infection, drug treatment, and other impacts. The complex mechanics of this coordinated response in cells have improved through evolution. Reverse engineering approaches may help demystify the complex mechanisms of cellular responses to external stimuli. The genome of every organism can be defined by a network of transcriptional regulatory relationships. In such a network, a pair of genes in the genome are related if one gene serves a regulatory or controlling role relative to the second gene. For instance, a gene may regulate another gene by promoting or inhibiting its expression. A pair of genes in a genome may be related if they are co-regulated, Co-regulated genes are simultaneously promoted toward expression or inhibited from expression by the same regulatory agent. Because a genome may contain many thousands of genes, genomics research requires a wide array of analytical approaches to determine its transcriptional regulatory network.

In some cluster-based experiments, clusters of genes are seen to be expressed simultaneously in nearly identical patterns. In the absence of definitive regulatory causation evidence, clustered genes expressed in nearly identical patterns are said to be related through co-expression.

A simple graph can illustrate a relationship between two entities. The terms *graph* and *network* are often used interchangeably. Entities in a graph are shown as nodes and a relationship between two nodes is shown with a connecting line called an edge. A graph of nodes and edges may help depict gene regulation and, from a larger perspective, the mechanics of life sustaining functions within a cell. For example, a node of a graph can be said to represent a gene and an edge can be said to represent a relationship between two genes.

Figure 6.1 is a graph that illustrates a hypothetical five-gene network. Each gene is represented by a graph node. Each line or edge connecting genes has an arrowhead depicting the direction of regulatory effect. The edges in this graph are said to be directed. Each of the three directed edges represents a regulatory effect of one gene on another gene. Specifically, the graph shows that gene G1 regulates gene G2 and gene G2 regulates genes G3 and G4. Gene G5 is neither a regulatory gene nor target gene in this gene network.

Figure 6.2 illustrates a separate hypothetical five-gene network. Each gene is represented by a graph node. Since there is no arrowhead connecting two genes on lines or edges, there is no regulatory effect depicted. The edges in this graph are said to be undirected. Each of the two undirected edges represents a co-expression relation between the connected genes. The expression pattern between a pair of co-expressed genes is very similar. This is useful information even where there is no evidence of a regulatory effect one gene may have on the other. Specifically, the graph shows that genes G1 and G2 and genes G3 and G5 are co-expressed. Gene G4 has no co-expression relationship with any other gene in the network.

In spite of intense research to develop an accurate gene regulatory network prediction algorithm, the mechanisms remain poorly understood. Researchers employ a wide variety of data mining and machine learning algorithmic approaches, as well as massive parallelization on graphical processing units (GPU), cloud-based technologies and community approaches [246, 247]. In some GRN approaches, complexity increases exponentially with the number of genes. Fortunately, creative heuristics improve performance of the NP-hard aspects of network inference in many situations [71, 87, 132].

Computational tools and public access databases are essential for cost-efficient deciphering of gene regulatory mechanisms. Publicly accessible databases of well studied model organisms are available to the research community, as described in Chapters 1 and 5. While FlyBase [84] is the recognized authority on *Drosophila melanogaster* (fruit fly) electronic documents and information processing software, the Encyclopedia of DNA Elements of Model Organisms (modENCODE) [312] comprehensively annotates cellular products of model organisms, including fruit fly. modENCODE provides hundreds of fruit fly-related data sets of great interest to network inference researchers. These modENCODE data sets represent cellular activity across different fruit fly developmental stages and cell types. Such activity is useful for predicting stage-specific and tissue-specific gene expression and regulation.

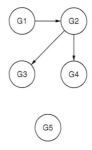

FIGURE 6.1: Hypothetical five-gene network showing that gene G1 has a regulatory effect on gene G2 and gene G2 has a regulatory effect on genes G3 and G4.

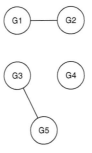

FIGURE 6.2: Another hypothetical five-gene network showing that genes G1 and G2 and genes G3 and G5 have co-expression relationships.

In this chapter, several tools for evaluating cellular products expressed under a host of different conditions are described. Each computational approach attempts to infer a gene regulatory network based on gene expression data. These algorithms use different computational approaches, are run on various software platforms, process both steady state and time series gene expression profile formats and apply different supervisory modes of operation.

Computational approaches to network inference

Computational biology analysis methods are critical for gleaning insights into the cellular mechanisms from massive numbers of gene expression datasets. In the field of computational biology, several approaches are used. There is little evidence that one approach offers or will offer the single best solution to the network inference puzzle. Rather, the researcher should explore the features of each and deploy them appropriately in the analysis of the gene expression datasets. Current well-established computational approaches include Bayesian networks, ordinary differential equations (ODEs), information theoretic and feature selection.

- *Bayesian network inference*
  A Bayesian network is a graphical model of probabilistic relationships among a set of random variables. In the case of a gene regulatory network, each variable in the graph represents one of the genes in the network. The Bayesian network inference approach is based on statistical dependencies [124, 293]. Each node of a Bayesian gene regulatory network represents the expression level of a gene and each edge represents a dependency between two genes. The expression level of a node, x_i on the graph, $G = (X, E)$, is dependent on the expression level of its parent node, $Pa(x_i)$, on the graph. The Bayesian network inference model is based on conditional probability, $P(x_i \mid Pa(x_i))$. If the probability of expression of each gene given its parent is known, the Bayesian network can be built. Note that there may be multiple parents for each node in the graph since the expression level of a gene may be regulated by more than one other gene.

  Bayesian network inference provides good prediction results for relatively small numbers of genes and special cases. However, inference using graphical networks is an NP-hard problem [71, 87, 132]. Therefore, for a network with a large number of genes, the compute time becomes prohibitive using Bayesian network inference.

  Bayesian-based approaches are commonly used in time series

data evaluation. Dynamic Bayesian networks (DBNs) allow for a feedback cycle in the network. Feedback cycles are not represented in standard Bayesian networks. Bayesian model averaging (BMA) approaches help reduce the effects of model uncertainty. In general, Bayesian-based approaches are computationally intensive and do not scale well as the number of genes in the network grows.

- *Ordinary differential equation (ODE) approach*
  The ordinary differential equation (ODE) model consists of a tightly coupled system of non-linear, ordinary (not partial) differential equations. Each ODE represents one of the genes in the network. The unknown parameters of each equation are inferred from gene expression data. Unknown parameters indicate the rate and direction of transcription product mass flow as well as the strength of the interactions from gene to gene. Typically, the unknown parameters of a system of ODEs are resolved simultaneously. Unfortunately, to infer the network and determine all parameters simultaneously is a large-scale, iterative optimization problem that is very time consuming and generally cannot be completed in an acceptable timeframe for a large gene network.

  A strategy to divide the problem into N sub-problems, i.e., an independent ODE for each of N genes, is promising [245]. In this strategy, the unknown parameters for each separate ODE are resolved independently of the other ODE equations. This approach lends itself well to a parallel processing technique presented in Chapter 7.

- *Information theoretic approach*
  The mutual information (MI) approach to the problem of reverse engineering a transcriptional regulatory network from an expression profile dataset involves identifying pairwise dependencies among all gene transcripts. In general, this is a problem of identifying dependencies between a pair of random variables [356]. Mutual information (MI) approaches help identify non-linear dependencies [369].

  By applying information theory, the MI between two random variables quantifies how closely the two variables are interrelated [330]. To compute MI, the expression rate of each gene over periodic time intervals is considered to be a random variable, e.g., X. Whether the gene's expression environment is in steady state or perturbed, the collection of the gene's expressions over time t is considered to represent x_t of the random variable X. When the expression pattern of one gene, i.e., random variable X, is very similar to the expression pattern of another gene, i.e., random variable Y, the mutual information between these genes, or random variables, indicate that two genes are closely interrelated.

When the expression patterns of two genes are similar and synchronized, the genes may be in a co-expression relationship, and their synchronized expression is possibly caused by the expression product of another gene. When the expression patterns of two genes are similar but not synchronized, the genes may be in regulatory relationship, and the expression pattern of one gene, after a delay, is reflected in the expression pattern of the second gene. By evaluating the mutual information among all pairs of genes in a gene expression dataset, it may be possible to infer the gene regulatory network. Tools using the mutual information approach for network inference are described later in this chapter.

- *Feature selection approach*
  The feature selection approach to the GRN inference problem leverages attribute knowledge (features) about two genes. A significance score is computed for each pair of genes based on features of each gene. If the computed significance score for a pair of genes, $G1$ and $G2$, exceeds a cutoff threshold, then $G1$ is predicted to be a regulator of gene $G2$. This approach can take into consideration, for instance, that a gene is known not to be a transcription factor (TF) gene, thus reducing the search space and improving compute time performance.

  Unfortunately, a gene feature that may be relevant in determining a regulatory relationship between two genes may not be obvious. The process for identifying relevant features is complex. Feature selection research is active in many areas of data processing including gene array expression analysis [101, 149, 181, 313].

Network inference software development platforms

Interestingly, while the costs of sequencing and biological data storage continue to fall, the data analysis process costs remain high. Much of the data analysis cost is related to the time invested in evaluating existing software tools, developing new software tools and conducting reproducible analytical research. As quantity, quality, visualization and user-friendliness of computational biology tools increase, researchers take advantage of well established and feature-rich platforms. An environment to conduct analysis also serves as a repository for related experiment data and documentation and intermediate and final results. Currently, two of the most popular development and analysis platforms are R and MATLAB.

- *R language and environment*
  R is a well established open source language and environment for statistical computing and graphics [239] (*https://www.r-project.org*). The interactive command line nature of R makes it highly effective for learning and computational biology tool development. For developers who may prefer a GUI, RStudio provides an open source integrated development environment (IDE) for R (*https://www.rstudio.com*). The graphical interface of RStudio contains multiple viewing panes exposing various session aspects including user command line interaction, graphing and plotting results, a searchable command history and working directory file contents.

  For a well established repository of R computational biology related software, Bioconductor (*http://www.bioconductor.org*) is widely used. Bioconductor is an open source software project providing development packages for analysis and comprehension of high-throughput genomic data. Bioconductor is based primarily on the R programming language. The standardized format of a Bioconductor package includes source code, user documentation and a tutorial-like vignette depicting a step-by-step procedure that serves as a brief introduction for running the package.

  Some of the tools presented in this chapter were developed using R.

- *MATLAB ® development platform*
  MATLAB (*http://www.mathworks.com*) is another well established development platform in many scientific areas. MATLAB is widely used for computational biology tool development [325]. It is a proprietary product that provides numerous user-programmable mathematical and numerical methods. Some users not familiar with MATLAB report that the MATLAB scripting language is easier to learn than other operating system-dependent stand-alone applications such as R.

  Some of the tools presented in this chapter were developed using MATLAB.

Gene expressions in network inference research

Transcriptome profiling methods including microarray and RNA sequencing (RNA-seq) analyses are generating rapidly growing amounts of data for evaluation and research. Many of these gene expression profiles are available from public databases as described in Chapter 1. The inference of a gene regulatory network from a gene expression profile is an active and challenging research problem.

Network inference research involves analyzing gene expressions in different formats. In every living form, gene expression is the process by which a gene product, i.e., an RNA transcript, is formed using information encoded in a DNA gene. A gene product is either converted to a protein product or remains in the RNA state to serve another cellular function. Gene expression supports the biological machinery for life. By studying gene expression products in a reverse engineering approach, the process will become better understood.

Of course, not every protein product in a cell is a transcription factor (TF) intended for gene regulation. Proteins fill a wide variety of non-regulatory functions. Furthermore, if a TF gene is expressed, it is not a certainty that its transcript will be translated into a TF protein. When evaluating gene expression data, uncertainties such as these must be taken into account [123].

Gene expressions from growing numbers of species pour into digital repositories at ever-accelerating rates. The network inference researcher chooses from different gene expression categories depending on his or her specific focus. For one computational experiment, a small synthetic dataset is appropriate, while a separate experiment may call for transcriptomes taken from a genetically modified organism. Here are several gene expression categories from which a network inference researcher might choose.

- *Biologically prepared or synthetically prepared gene expression*
  Microarray and RNA-seq gene expression data represent gene expression products from a biological organism. Public databases of biological gene expression data, especially GEO [33, 107], are growing at an exponential rate due to advances in speed and availability of sequencing devices. While most of electronic gene expression datasets are generated from biological organisms, synthetically generated gene expression datasets are available and offer several distinct advantages over their biologically generated counterparts.

  The simulation of *in silico* networks and associated gene expression datasets have the advantages of being fast to create, easy to reproduce and less expensive to produce compared with biological experiments. A synthetic gene expression dataset may be produced using open source tools [322] that generate *in silico* gene networks modeled after biological networks of well understood organisms like *S. cerevisiae* [140] and *E. coli* [44]. A smaller gene network subset of any size can be extracted readily from a full sized *in silico* gene network. A simulated gene expression, which is limited to the number of genes in the simulated gene network, is produced by the software. This capability allows a network inference developer the flexibility to conduct meaningful tests on gene expressions consisting of 10 or 100 genes rather than the 4000+ or 6000+ genes in the full *E. coli* and *S. cerevisiae* gene networks, respectively.

The DREAM series competitions used the smaller gene network development approach with successful results [247]. The dialogue for reverse engineering assessments and methods (DREAM) challenges were established 2007 to help promote the advancement of GRN capability through collaboration. Synthetically prepared gene expression datasets used in these competitions provide developers much improved turnaround times due to the more manageable dataset sizes. These datasets are software generated and based on the GRN computer graph model derived using gene regulatory knowledge acquired from model organisms. This GRN computer graph model serves as the gold standard reference to help determine the best inferred network solution for the DREAM problem posed.

To expedite algorithm development, small gene regulatory networks can be defined and solved using gene net weavers to establish small synthetic datasets. An algorithm developer will start testing with a dataset of only a handful of genes and grow the network for more comprehensive testing as the algorithm develops. To help expedite GRN prediction algorithm development, smaller yeast GRN computer graph models are used. The smaller model is created by choosing a subset of genes from the yeast genome. This subset represents the vertices of the GRN computer graph model and all known interactions between pairs of genes in this subset represent the graph edges. GRN prediction algorithm development can be substantially expedited using smaller GRN computer graph models due to correspondingly shorter runtimes.

- *Knockdown or knockout gene expression*
  The role or activity of a gene product is sometimes identified through the use of reverse genetics techniques [236], such as gene knockdown and gene knockout. In reverse genetics, a functional mutation is produced by removing a gene of interest from the DNA. Alternatively, gene expression is affected by introducing an agent to limit protein translation. In a gene knockdown experiment, the expression of a gene of interest is partially reduced. In the case of

gene knockout, the gene expression is eliminated completely. By studying cellular gene expression dynamics where the effect of a specific gene is reduced or eliminated, much can be learned about the role of that gene. Cellular response to knockdown and knock-out reverse genetics provides valuable clues for reconstructing the regulatory transcriptional network. Determining relative regulatory roles of genes can be aided by using knockdown and knockout techniques as follows.

The expression capability of a specific gene or several specific genes can be partially or fully restricted in the DNA. A partial expression restriction of a gene is known as a knockdown. The full expression restriction of a gene is known as a knockout. Where no genes are knocked down or knocked out, the genome is said to be in its wild type (WT) state. In the presence of knocked down or knocked out genes, a steady state expression profile will show compensatory expression values among the remaining genes. This response provides valuable clues for reconstructing the regulatory transcriptional network.

- *Wild-type gene expression*
  Wild-type refers to the phenotype of the typical form of a species as it occurs in nature. A wild-type organism serves as the original strain prior to the introduction of a mutation for research. In network inference research, wild-type gene expression is used as a reference in comparing naturally occurring genotypes against those of the mutated counterparts.

- *Steady state versus time series gene expression*
  A steady state gene expression dataset represents transcription levels during a disturbance-free or perturbation-free condition. Transcript levels reflect routine cellular activity to sustain normal life functions. Where no genes are knocked down or knocked out, the genome is said to be in its wild-type state. In the presence of a knocked down or a knocked out gene, a steady state expression profile will show compensatory expression values among the remaining genes.

Interestingly, there is evidence of a set of "housekeeping" genes in a living organism [110]. These genes are expressed in all conditions in all tissues. It is believed that cellular mechanisms regulate housekeeping genes so that internal conditions remain stable and relatively constant. The steady state response to a knocked down or knocked out gene provides valuable clues about how certain housekeeping genes compensate for housekeeping genes that are reduced or missing altogether.

A time series gene expression dataset is created when a specimen sample is deliberately disturbed or perturbed from its steady state by the introduction of a physical or chemical environmental change intended to invoke a gene expression response. Following a controlled perturbation and at specific time intervals a gene expression "snapshot" is recorded and gathered in a time series analysis dataset for evaluation. In some experiments, the gathering of time series gene expression snapshots will continue until the cellular steady state is restored. In addition, observing time series gene expression should provide a greater understanding of many complex biological processes, such as disease progression, development and drug responses as each step of a process is watched [19]. While a growing range of biological systems lend themselves to time series expression experiments, challenges must be overcome before these studies accurately predict a biological system [31]. For instance, oversampling may improve the overall observation of a system, but will create budgetary constraints due to the cost. Conversely, undersampling runs the risk of missing important gene interactions.

Time series expression evaluation may provide an opportunity to reverse engineer the mechanisms that regulate responses. Questions that remain unanswered relate to biological systems analysis (e.g., cell life cycles), response control (e.g., medication dosage), normal cell development, and diseased cell development.

Conversely, an invalid sample called an unlabeled data item is expected to be provably false. Unfortunately, an unlabeled data item is more difficult to prove. For instance, in the presence of three separate conditions, a claim may be made that a specific result is guaranteed not to occur. The problem is that by adding a fourth, as yet unknown, condition, the specific result may indeed occur and be contradictory to the initial claim.

An introduction to classification experiments using labeled and unlabeled data is presented in Section 6.2.2.1 of this chapter.

Algorithmic supervision in network inference

Network inference algorithms are designed as unsupervised, supervised or semi-supervised [244]. The level of supervision employed by an algorithm depends on the use of an inference guiding model. This concept is similar to genome sequence mapping (Chapter 7) using a *de novo* approach as compared to a reference-based reconstruction of an original genome sequence [273].

Supervised or model-based network inference algorithms achieve the highest accuracy among all the network inference algorithms. A supervised algorithm requires a set of training data consisting of large numbers of positive and negative samples. Training data are used to build or train a classification model. A classification model or classifier is built after statistical analysis of available positive samples. Depending on the amount of research performed on an organism, availability of positive samples varies. For instance, much more information is available about well studied model organisms like yeasts and bacteria. Network inference prediction accuracy improves as the number of training samples for the classifier increases.

When gathering samples to create a supervised algorithm classifier model, a positive sample represents a known gene regulatory effect. A positive sample has been proven in a laboratory experiment. Conversely, a negative sample should represent a gene regulatory effect that is known to never occur. Proving the absence of a gene regulatory effect is more challenging than proving a positive gene regulatory effect. Researchers typically treat a negative sample as potentially non-conclusive. Experimental results from studies using negative samples reflect the understanding that a negative sample may not be as conclusive as a positive sample. A negative sample provides value in supervised and semi-supervised studies as described in this chapter. Researchers may employ an unknown gene interaction as a negative sample in a supervised or semi-supervised study [70].

Unsupervised or model-free algorithms infer networks based solely on gene expression profiles and do not use training samples. Model-free methods are easily scalable but generally suffer from a lack of interpretability. The accuracy of unsupervised algorithms is usually low. However, unsupervised algorithms are useful in situations where training data are not available for an organism being studied. Semi-supervised algorithms often exploit positive unlabeled (PU) learning techniques by taking a small sample of positive examples and a large number of unlabeled examples to train a classification model and use the trained model to predict a network.

Network inference tools

Biological network inference tools for steady state expression data presented in this chapter are the following:

- *ARACNE*

  The Algorithm for the Reconstruction of an Accurate Cell Network (ARACNE) is an unsupervised bioinformatics tool that uses an information theoretic approach to infer a gene regulatory network (GRN) [250]. ARACNE evaluates steady state gene expression profile data and predicts the regulatory relationships of genes based on mutual information (MI) computed between gene pairs. A novel technique called Data Processing Inequality (DPI) is used in ARACNE to reduce the number of false positive predictions. In the process of computing the MI between two genes, there are many situations where two genes appear to be directly related when in fact there is a third gene involved. Using DPI, ARACNE eliminates such indirect relationships where a gene is related to a second gene but only indirectly through a third gene. With DPI, the GRN edges predicted by ARACNE are more accurate than those predicted by other algorithms using the information theoretic approach.

  ARACNE is available as an R package and can be downloaded freely from Bioconductor (*http://www.bioconductor.org*).

- *CLR*

  Context Likelihood of Relatedness (CLR) is an unsupervised algorithm that uses an information theoretic approach [261, 330] to predict network graph edges of a gene regulatory network (GRN) [115]. CLR evaluates steady state gene expression data and predicts the regulatory relationships of genes based on mutual information (MI) computed between gene pairs. Like ARACNE, CLR applies a corrective step to reduce false predictions.

  Mutual information (MI) scores are computed for all possible pairs of transcription factor (TF) genes with potential target genes (TGs). Then, for each pair of TFs and TGs, CLR compares the TF-TG score with the scores of every other pair in which the TF or TG is represented. This process removes lower scoring pairs that may be misinterpreted as significant regulatory pair predictions. By applying this corrective step, CLR's prediction results are more accurate when compared with other GRN inference algorithms that use the information theoretic approach. Furthermore, the CLR algorithm was among the fastest of GRN inference algorithms in an independent benchmark test [36].

  CLR is available as an R package and can be downloaded freely from Bioconductor (*http://www.bioconductor.org*).

- *GENIE3*

  The GEne Network Inference with Ensemble of trees (GENIE3)

algorithm [181] uses the feature selection computational approach. Developed as a MATLAB application, GENIE3 evaluates steady state gene expression profile data and predicts the regulatory relationships among p genes by decomposing the prediction problem into p separate and independent regression problems. GENIE3 applies tree ensemble methods [50, 135] for efficient prediction of high ranking target genes with respect to each potential regulator gene candidate. In each of the regression problem for a target gene, the expression pattern of the target gene can be predicted based on the expression patterns of all transcription factors. The significance of each transcription factor in the prediction of the target gene determines the apparent regulatory relationship.

GENIE3 was the best performing GRN inference algorithm in the DREAM4 challenge in 2009 [247].

- *MRNET*
  The Minimum Redundancy NETworks (MRNET) unsupervised algorithm [260, 261] infers a gene regulatory network (GRN) from steady state gene expression profile data. MRNET deciphers GRNs using a feature selection technique known as Minimum Redundancy Maximum Relevance (MRMR) [101]. The MRMR technique utilizes an information theoretic computational approach. The MRNET algorithm performs a feature selection for each candidate target gene based on the expression of all other genes (i.e., candidate regulator genes). A feature-based ranked list maximizes the mutual information (MI) between potential regulator and target genes (maximum relevance) while simultaneously ensuring mutual dissimilarity among candidate regulator genes (minimum redundancy).

  MRNET is available as an R package and can be downloaded freely from Bioconductor (*http://www.bioconductor.org*).

Biological network inference tools for time series expression data presented in this chapter are the following:

- *Inferelator*
  The Inferelator unsupervised GRN inference algorithm [47, 243] makes predictions by solving a system of ordinary differential equations (ODEs) given gene expression (both time-series and steady state) profiles and prior relevant knowledge. Inferelator can be run using a Web based server or by downloading and executing open source R software.

  Inferelator was used to demonstrate that strengths and weaknesses among different GRN inference methods can be leveraged in combination to achieve improved results. Inferelator is a three-component GRN inference pipeline process which was awarded a tie for first among 19 methods in the DREAM4 100-gene *in silico* network inference challenge in 2009 [141].

- *Jump3*
  Jump3 is an unsupervised GRN inference algorithm [135, 182] that uses the feature selection computational approach. Developed as a MATLAB application, Jump3 evaluates time series gene expression profile data and predicts the regulatory gene parings. Jump3 deploys the on/off model of gene expression [299] to interpret the transcription level of a target gene as falling into on or off discretized levels, depending on the state of its promotor. Like GE-NIE3 [181], Jump3 applies tree-based ensemble methods [50, 135] to help bridge the gap between model-based and model-free computational approaches to GRN inference.

- *ScanBMA*
  ScanBMA is an unsupervised GRN inference algorithm [413] that uses the Bayesian network inference computational approach and was developed within the R language and environment . ScanBMA reads time series gene expression profile data and predicts regulatory relationships between pairs of genes in a GRN. ScanBMA addresses known limitations of Bayesian network inference, such as the exponential computational costs of evaluating networks that are not small [71, 87]. A novel Bayesian model averaging (BMA) approach utilizes a more efficient model space search. ScanBMA also performs time series expression profile data transformations to improve performance in several areas, such as eliminating predictions of gene self-regulation.

  ScanBMA is available as part of the NetworkBMA R package, which can be downloaded freely from Bioconductor (*http://www.bioconductor.org*).

- *TimeDelay-ARACNE*
  TimeDelay-ARACNE applies the ARACNE GRN inference algorithm to time series gene expression profile data [422], whereas ARACNE is intended for the analysis of steady state gene expression profile data. As a descendant of ARACNE, TimeDelay-ARACNE is an unsupervised algorithm and uses the information theoretic computational approach. Like ARACNE, TimeDelay-ARACNE was developed with the R language and environment.

  TimeDelay-ARACNE performs three novel processing steps on

a time series gene expression profile dataset and generates a GRN-directed graph. First, TimeDelay-ARACNE identifies the initial change of expression (IcE) for each gene across all time data points. IcE identifies when a gene's expression value first exceeds a defined threshold and helps reveal a target of a regulator gene with an earlier IcE. Next, a network is constructed by computing mutual information (MI) scores for each potential regulator-target gene pair and for multiple time shifts based on IcE within each gene pair. Finally, TimeDelay-ARACNE prunes the constructed network, as in ARACNE, using a technique called Data Processing Inequality (DPI). DPI pruning reduces the number of false positive predictions.

The TimeDelay-ARACNE R package, named TDARACNE, can be downloaded freely from the Bioconductor software project (*http://www.bioconductor.org*).

Performance of an algorithm is measured by its ability to produce a correct result in an acceptable period of time, i.e., its time complexity. Most graph-based algorithms perform well processing a small number of vertices. However, the scalability of an algorithm is measured by its ability to process increasing amounts of data, e.g., a larger gene network, in an acceptable time, which is an important research concern. The time complexity of a GRN prediction algorithm also impacts algorithm development time when an algorithm does not scale well with increasingly large gene networks.

In addition to time efficiency, the accuracy of a GRN prediction algorithm is important. Algorithm accuracy can be measured using a confusion matrix, which presents key algorithm metrics. For one example, specificity is a measure of incorrect prediction, i.e., the number of predictions that are incorrect. As another example, recall is a measure of prediction completeness, i.e., the number predicted versus the actual number. Confusion matrix metrics are valuable in determining the accuracy of an algorithm. However, they can only be computed when there is truth model, or GRN gold standard, to compare predictions against.

6.2.1 Unsupervised Inference Methods

An unsupervised algorithm infers a network *de novo*, i.e., based solely on gene expression profiles. No training samples are available for reference when making a prediction. The accuracy of unsupervised algorithms is typically lower than that of supervised or semi-supervised approaches. However, unsupervised algorithms are useful where training data are not available for the subject studied.

To help advance the quality of unsupervised inference methods through collaboration, the dialogue for reverse engineering assessments and methods (DREAM) challenges were established in 2007 [247]. In each challenge, teams compete to devise the best solution to a complex problem. In one of these challenges, the DREAM4 challenge in 2009, problems to be solved fell into two categories: cellular network inference and quantitative model building. For the DREAM4 cellular network inference challenge, gene expression profile datasets were synthetically created using GeneNetWeaver [322]. Synthetic gene expression profiles datasets were based on known behaviors of well-studied model organisms, such as *Saccharomyces cerevisiae* [140] (yeast) and *Escherichia coli* (bacteria) [44]. *S. cerevisiae's* 6,000+ genes' properties and behavior under numerous conditions have been extensively studied. Much is known from wet lab experiments and graph-based computational modeling about how these thousands of genes interact. A full yeast GRN computer graph model consists of 6,000+ vertices representing genes and thousands of edges representing known interactions between pairs of genes.

Datasets from the DREAM4 challenge (*http://dreamchallenges.org/*) were prepared by first extracting 10-gene and 100-gene subnetworks from known regulatory networks of *E. coli* and *S. cerevisiae*. Known self-regulatory interactions were removed from each subnetwork. Therefore, no self-regulating edge appeared in the GRN predicted from each method. Each dataset represented synthetic messengerRNA (mRNA) concentration levels of each respective gene of the 10-gene or 100-gene subnetwork. The mRNA concentration levels were based on statistical models of the dynamics of each network. Synthetic DREAM4 datasets included steady state and time series configurations. The 10-gene and 100-gene networks and synthetic gene expression profile data were generated using GeneNetWeaver (GNW) [322] (*http://gnw.sourceforge.net*). Each 10-gene and 100-gene subnetwork served as the gold standard truth reference against which reverse engineered predictions were compared for accuracy.

Presented here are simple experimental results from running four unsupervised GRN inference algorithms using steady state expression data and four unsupervised GRN inference algorithms using time series expression data. For each of the steady state data experiments, the same input data file was evaluated, and for each of the time series data experiments, the same input data file was evaluated. However, in all eight experiments, the input data files were synthetically derived from the same 10-gene network for which the gold-standard reference network was known (cf. Table 6.1 and Figure 6.3). Simulated mRNA gene expression levels are based on a gold standard network. Two critical metrics

TABLE 6.1: Gold standard regulatory gene pairs for DREAM4 10-gene network 1

Regulator Gene	Target Gene
G1	G2
G1	G3
G1	G4
G1	G5
G3	G4
G3	G7
G4	G3
G6	G2
G7	G3
G7	G4
G8	G2
G8	G6
G9	G10
G10	G3
G10	G4

that determine GRN method prediction accuracy are precision and recall. The precision metric measures the ability of the GRN method to accurately predict whether a gene regulates another gene. Precision is computed as the number of true positive (TP) predictions divided by the sum of all predictions, i.e., true positives and false positives (FPs), i.e., $TP/(TP + FP)$. The recall metric measures the percentage known interactions predicted by the GRN method. Recall is computed as the number of true positives divided by the sum of true positives and false negatives (FN), i.e., $TP/(TP + FN)$.

Cytoscape is used for visual assistance when arranging the graph nodes for optimal appearance, i.e., with minimal crossing edges [314] (*http://www.cytoscape.org*). Cytoscape's visualization tool allows customization of a graph layout to help improve figure readability. Once an ideal graph layout has been designed using Cytoscape, x and y planar coordinates are used in a Perl script to generate postscript for each graph figure.

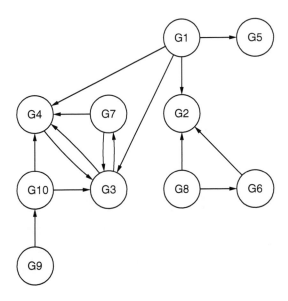

FIGURE 6.3: Gold standard-directed graph for DREAM4 10-gene network 1.

6.2.1.1 Steady-state Data Methods

Four unsupervised GRN inference steady state methods were evaluated in separate experiments. The input file for all four steady state experiments was from the DREAM4 challenge (*http://dreamchallenges.org/*). In each experiment, the same 10-gene DREAM4 steady state gene expression profile data was processed. The results of each experiment were presented in the form of a listing of predicted regulatory relationships and a directed or undirected graph depending on the results of each algorithm. For each of the steady state data experiments, the same 10-gene DREAM4 steady state input data file was used. The steady state dataset chosen was a 10-gene knockdown dataset (cf. Table 6.2).

ARACNE method

The ARACNE (Algorithm for the Reconstruction of an Accurate Cell Network) [250] R package is free to be downloaded from Bioconductor (*http://www.bioconductor.org*) and run against a steady state gene expression dataset to predict edges of a gene regulatory network (GRN). ARACNE is included in a suite of information theoretic-based GRN inference algorithms in an R package called Mutual Information NETwork inference (MINET) [261]. The steady state dataset chosen was the DREAM [247] synthetic 10-gene knockdown dataset (cf. Table 6.2). The ARACNE R package was run using the RStudio integrated development environment for R (*https://www.rstudio.com*). ARACNE run instructions are presented in a vignette with the ARACNE R package. Results of running ARACNE are listed in Table 6.3. The top ten edges predicted by ARACNE are shown as an undirected graph in Figure 6.4.

CLR method

The Context Likelihood of Relatedness (CLR) inference algorithm [115] R package is freely downloadable from the Bioconductor software project website (*http://www.bioconductor.org*). Like ARACNE, CLR is included in a suite of information theoretic-based GRN inference algorithms in an R package called Mutual Information NETwork inference (MINET) [261].

TABLE 6.2: DREAM4 synthetic knockdown gene expression profile

G1	G2	G3	G4	G5	G6	G7	G8	G9	G10
0.389841	0.072233	0.347800	0.639695	0.345609	0.316064	0.624457	0.649230	0.666183	0.692366
0.835776	0.035775	0.412133	0.660745	0.148526	0.332636	0.463641	0.816428	0.776379	0.584134
0.681233	0.084093	0.264790	0.706749	0.165805	0.256804	0.535463	0.705517	0.684542	0.779180
0.680201	0.060726	0.411928	0.446698	0.074264	0.290092	0.475431	0.599874	0.677533	0.778699
0.636300	0.075023	0.410319	0.721010	0.122417	0.253224	0.478014	0.618509	0.657348	0.690829
0.674216	0.115066	0.324927	0.652347	0.154404	0.163025	0.596110	0.711288	0.795991	0.698557
0.770595	0.116364	0.248213	0.521838	0.109264	0.274521	0.358635	0.639091	0.655159	0.683031
0.816394	0.178358	0.501203	0.676508	0.100178	0.521625	0.413467	0.321702	0.665357	0.673437
0.747799	0.108989	0.413700	0.715692	0.115321	0.227635	0.601653	0.749398	0.415616	0.620812
0.689105	0.138617	0.453761	0.747617	0.082723	0.300450	0.638665	0.691092	0.811923	0.264575

TABLE 6.3: ARACNE result after processing DREAM4 knockdown gene expression profile of *E. coli*

	G1	G2	G3	G4	G5	G6	G7	G8	G9	G10
G1	0	0	0	0	0	0.3352248	0.5996104	0	0	0.8188043
G2	0	0	0	0.1899108	0.2216251	0	0	0	0	0
G3	0	0	0	0	0.5680967	0.4283425	0	0	0	1
G4	0	0.1899108	0	0	0	0	0.3798874	0	0	0.4283425
G5	0	0.2216251	0.5680967	0	0	0	0	0.4808811	0	0
G6	0.3352248	0	0.4283425	0	0	0	0	0.1751170	0	0
G7	0.5996104	0	0	0.3798874	0	0	0	0.2747984	0.2216251	0
G8	0	0	0	0	0.4808811	0.1751170	0.2747984	0	0	0
G9	0	0	0	0	0	0	0.2216251	0	0	0
G10	0.8188043	0	1	0.4283425	0	0	0	0	0	0

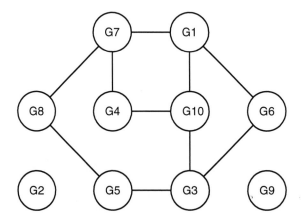

FIGURE 6.4: Top 10 GRN edges as predicted by ARACNE algorithm.

TABLE 6.4: Ranking of top 10 gene regulation pairs predicted by CLR algorithm for DREAM4 knockdown network 1

Regulator Gene	Target Gene	Predicted Probability
G3	G10	1.000000000000000
G5	G8	0.930145442485809
G1	G10	0.849079608917236
G1	G7	0.834647357463837
G7	G9	0.717656135559082
G3	G5	0.683408081531525
G3	G6	0.652681708335876
G4	G7	0.521122753620148
G8	G9	0.516775846481323
G2	G5	0.516481280326843

CLR was run against a steady state gene expression dataset to predict edges of a gene regulatory network (GRN). The steady state dataset chosen was the DREAM4 [247] synthetic 10-gene knockdown dataset (cf. Table 6.2). The CLR R package was run using the RStudio integrated development environment for R (*https://www.rstudio.com*). CLR run instructions are presented in a vignette with the MINET R package documentation. Results of running CLR are listed in Table 6.4. The top ten edges predicted by CLR are shown as a directed graph in Figure 6.5.

GENIE3 method

The GEne Network Inference with Ensemble of trees (GENIE3) algorithm [181] uses the feature selection computational approach with tree ensemble methods [50, 135]. The GENIE3 MATLAB application was downloaded (*http://www.montefiore.ulg.ac.be/∼huynh-thu*) and run against a steady state gene expression dataset to predict the edges of a 10-gene regulatory network (GRN). The steady state dataset chosen was the DREAM4 [247] synthetic 10-gene knockdown dataset (cf. Table 6.2). Results of running GENIE3 are listed in Table 6.5. The top ten edges predicted by GENIE3 are shown as a directed graph in Figure 6.6.

MRNET method

The Minimum Redundancy NETworks (MRNET) unsupervised algorithm [260] R package is freely downloadable from the Bioconductor software project website (*http://www.bioconductor.org*). Like ARACNE

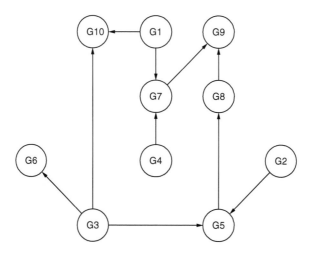

FIGURE 6.5: Top 10 GRN edges as predicted by CLR algorithm.

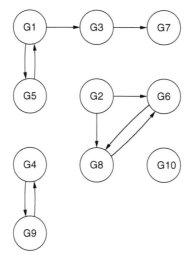

FIGURE 6.6: Top 10 GRN edges as predicted by GENIE3 algorithm.

TABLE 6.5: Ranking of top 10 gene regulation pairs predicted by GE-NIE3 algorithm for DREAM4 knockdown network 1

Regulator Gene	Target Gene	Predicted Probability
G4	G9	0.230069668336489
G1	G5	0.225208954726292
G5	G1	0.220131728788448
G3	G7	0.211605105383277
G8	G6	0.210537409550244
G9	G4	0.205067326708992
G6	G8	0.197078777480003
G2	G8	0.192526319994338
G2	G6	0.189476027305577
G1	G3	0.172723024331032

and CLR, MRNET is included in a suite of information theoretic-based GRN inference algorithms in an R package called Mutual Information NETwork inference (MINET) [261].

MRNET was run against a steady state gene expression dataset to predict edges of a gene regulatory network (GRN). The steady state dataset chosen to be processed was the DREAM4 [247] synthetic 10 gene knockdown dataset (cf. Table 6.2). The MRNET R package was run using the RStudio integrated development environment for R (*https://www.rstudio.com*). MRNET run instructions are presented in a vignette within the R package documentation. Results of running MR-NET are listed in Table 6.6. The top ten edges predicted by MRNET are shown as a directed graph in Figure 6.7.

6.2.1.2 Time Series Data Methods

Four unsupervised GRN inference time series methods were evaluated in separate experiments. The input file for all four time series experiments was from the DREAM4 challenge (*http://dreamchallenges.org/*). In each experiment, the same 10-gene DREAM4 time series gene expression profile dataset was processed. The results of each experiment were presented in the form of a listing of predicted regulatory relationships and a directed graph. Each method was applied to the same time gene expression profile dataset, which is the first of the five datasets produced for the DREAM4 10-gene network (cf. Table 6.7).

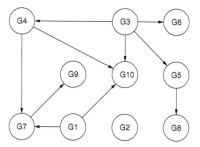

FIGURE 6.7: Top 10 GRN edges as predicted by MRNET algorithm.

TABLE 6.6: Ranking of top 10 gene regulation pairs predicted by MR-NET algorithm for DREAM4 knockdown network 1

Regulator Gene	Target Gene	Predicted Probability
G3	G10	1.000000000000000
G1	G10	0.818804383277893
G1	G7	0.599610507488251
G3	G5	0.568096816539764
G5	G8	0.480881154537201
G4	G10	0.428342580795288
G3	G6	0.428342580795288
G4	G7	0.374691098928452
G3	G4	0.247149929404259
G7	G9	0.221625164151192

Inferelator method

The Inferelator unsupervised GRN inference algorithm [47, 243] makes predictions by solving a system of ordinary differential equations (ODEs). Inferelator can be run by downloading and executing open source R software (*http://err.bio.nyu.edu/inferelator/*) or by using a Web-based server application (*http://dream.broadinstitute.org/gp/*).

The Inferelator Web-based server application was used to predict the GRN of the first of five 10-gene DREAM4 challenge [247] synthetically generated networks. Table 6.7 shows the input data (first dataset of five) processed by Inferelator. Table 6.8 lists the top ranked edges predicted by Inferelator. Figure 6.8 shows a directed graph of the top ranked edges predicted by Inferelator. The Inferelator top ranked edges listed and illustrated are edges whose predicted probability exceeded 40%.

Jump3 method

Jump3 is an unsupervised MATLAB GRN inference algorithm [135, 182] that uses the feature selection computational approach. The MATLAB implementation of Jump3 is available for public download (*http://homepages.inf.ed.ac.uk/vhuynht/*).

TABLE 6.7: DREAM4 synthetic time series gene expression profile dataset

Time Point	G1	G2	G3	G4	G5	G6	G7	G8	G9	G10
0	0.6665114	0.1272186	0.3550646	0.7745716	0.1004299	0.275493	0.6067846	0.7430983	0.6656366	0.6950638
50	0.3257748	0.1218223	0.3464115	0.7229108	0.1924591	0.3107637	0.6096963	0.7567515	0.5554138	0.7327167
100	0.1775012	0.0443587	0.5712888	0.586828	0.2333497	0.3569736	0.4647324	0.6656993	0.7211032	0.6717156
150	0.1838851	0.0615354	0.4849771	0.6338205	0.40451	0.3375671	0.4200074	0.7324089	0.6402268	0.6970434
200	0.0930693	0.1398431	0.3435551	0.5354375	0.5583316	0.2994227	0.4145393	0.6771472	0.5478739	0.7512751
0	0.6927804	0.1735119	0.3080785	0.5990709	0.1445523	0.2494241	0.5905584	0.6367970	0.6891143	0.6944974
50	0.6137000	0.0954328	0.2815777	0.6576991	0.2136285	0.3952169	0.2199619	0.5711943	0.6843410	0.6816443
100	0.5923916	0.0704157	0.2896261	0.6349119	0.2556762	0.3336192	0.1706594	0.7645988	0.6408686	0.6201303
150	0.5868361	0.0880437	0.1935983	0.5846269	0.1315485	0.3849495	0.1213159	0.7072442	0.8166294	0.7047542
200	0.6293309	0.1149911	0.2186433	0.4509727	0.1687147	0.3736593	0.0786808	0.6970827	0.8038210	0.6925731
0	0.7025546	0.0965407	0.4955336	0.6290043	0.1394119	0.2777284	0.5852446	0.6987308	0.7189792	0.6317014
50	0.2670111	0.0066399	0.4120065	0.6331020	0.1601289	0.3160889	0.3879914	0.3125419	0.8405381	0.7525926
100	0.1495577	0.0002031	0.3162661	0.6388113	0.3527526	0.4121452	0.5371867	0.2207124	0.7847303	0.6308077
150	0.1258573	0.0000726	0.3444491	0.5332911	0.5828865	0.4561672	0.5854555	0.1641590	0.6864760	0.6603441
200	0.1503708	0.0007181	0.3601180	0.4870183	0.6474004	0.5160537	0.4715633	0.1138825	0.7514894	0.7220521
0	0.7291507	0.1646974	0.5393216	0.7087402	0.1237001	0.2432843	0.3970339	0.7599215	0.6285854	0.6076601
50	0.5109302	0.1803674	0.3894485	0.6874071	0.1012508	0.4088945	0.1468781	0.3528068	0.4169359	0.7657149
100	0.6097374	0.1875477	0.2565762	0.5980135	0.1486603	0.4252275	0.0264118	0.3565534	0.4086711	0.8156564
150	0.7059362	0.2070564	0.1696391	0.4758323	0.1471183	0.4733397	0.0259651	0.2819844	0.3558222	0.7042378
200	0.7791460	0.2695352	0.2447503	0.3854643	0.1409376	0.5552323	0.0067659	0.1862479	0.4058957	0.6916045
0	0.7147163	0.0789012	0.4215624	0.6301302	0.0850342	0.3092021	0.4516194	0.6951692	0.8362663	0.6439763
50	0.6338617	0.1135145	0.2933042	0.6494650	0.0442124	0.3514063	0.2277307	0.7833732	0.3862019	0.6660204
100	0.6247811	0.0807247	0.3235387	0.5906159	0.0269790	0.3053646	0.1633944	0.8113616	0.1997879	0.6863509
150	0.7196440	0.0561945	0.3469269	0.4836713	0.0273254	0.3056355	0.1746293	0.8313165	0.1924133	0.7170860
200	0.7094180	0.0723323	0.4130727	0.6042179	0.0469687	0.3537859	0.1830969	0.6419759	0.0529535	0.5236239

NOTE: The gene expressions in this dataset are from the first of five DREAM4 10-gene networks. Shown are the first 5 of 21 observations from each of the five time series in the dataset.³

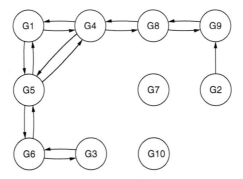

FIGURE 6.8: GRN edges as predicted by Inferelator algorithm.

TABLE 6.8: Ranking of gene regulation pairs predicted by Inferelator algorithm for DREAM4 time series network 1

Regulator Gene	Target Gene	Predicted Probability
G4	G1	1.000000000000000
G5	G1	0.903008447772773
G1	G4	0.889178747034423
G4	G5	0.853247573645150
G8	G4	0.840557755418847
G1	G5	0.805023200241754
G5	G4	0.613264848869402
G4	G8	0.499993018826361
G5	G6	0.490255264686652
G6	G5	0.490255264686652
G2	G9	0.457091312465209
G8	G9	0.448451035445612
G9	G8	0.448451035445612
G3	G6	0.445366110691197
G6	G3	0.445366110691197

NOTE: Shown are pairs whose predicted probability exceeds 40%.

Jump3 was run to predict the GRN of the first of five 10-gene time series DREAM4 challenge [247] synthetically generated networks. Table 6.7 shows the input data processed by Jump3. Table 6.9 lists the top ranked edges predicted. Figure 6.9 shows a directed graph illustration of the top ranked edges predicted by Jump3. The Jump3 top ranked edges listed and illustrated are those whose predicted probability exceeds 20%.

ScanBMA method

ScanBMA is an unsupervised GRN inference algorithm [413] that uses the Bayesian network inference computational approach. ScanBMA was developed within the R language and environment and is downloadable from Bioconductor as part of the NetworkBMA R package (*http://www.bioconductor.org*).

ScanBMA was downloaded and run in the RStudio environment (*https://www.rstudio.com*) using the first of five DREAM4 synthetic 10-gene time series profile datasets. Table 6.7 shows the input data (first data set) processed. Table 6.10 shows the top ranked edges predicted. Figure 6.10 shows the top ranked graph edges predicted by ScanBMA.

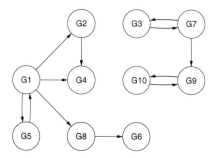

FIGURE 6.9: GRN edges as predicted by Jump3 algorithm.

TABLE 6.9: Ranking of gene regulation pairs predicted by Jump3 algorithm for DREAM4 time series network 1

Regulator Gene	Target Gene	Predicted Probability
G8	G6	0.886894
G3	G7	0.841794
G9	G10	0.831725
G1	G5	0.782503
G1	G2	0.715546
G7	G3	0.512323
G5	G1	0.479648
G1	G4	0.357726
G2	G4	0.236863
G1	G8	0.231089
G10	G9	0.228776
G7	G9	0.207139

NOTE: Shown are pairs whose probability exceeds 20%.

The ScanBMA top ranked edges listed and illustrated are edges whose predicted probability exceeds 90%.

TimeDelay-ARACNE method

TimeDelay-ARACNE applies the ARACNE GRN inference algorithm to time series gene expression profile data [422]. TimeDelay-ARACNE accepts a time series gene expression profile dataset and generates a GRN directed graph. TimeDelay-ARACNE is available for download as an R package from Bioconductor (*http://www.bioconductor.org*).

TimeDelay-ARACNE (TDARACNE) was downloaded and run in the RStudio environment (https://www.rstudio.com) using the first of five DREAM4 synthetic 10-gene time series profile datasets. Table 6.7 shows the input data (first data set) processed by TimeDelay-ARACNE. Table 6.11 shows the edges predicted and Figure 6.11 shows the regulatory graph edges predicted by TimeDelay-ARACNE.

6.2.2 Supervised and Semi-supervised Inference Methods

Use of computational methods to infer gene regulatory networks (GRNs) from gene expression profiles is a challenging task. Unsupervised

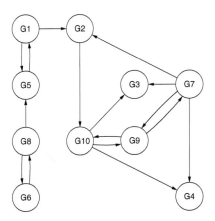

FIGURE 6.10: GRN edges as predicted by NetworkBMA algorithm.

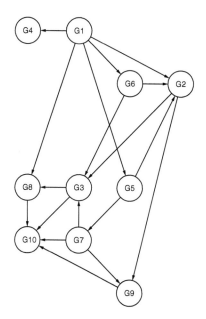

FIGURE 6.11: GRN edges as predicted by TimeDelay-ARACNE algorithm using DREAM4 10-gene dataset 1.

TABLE 6.10: Ranking of gene regulation pairs predicted by NetworkBMA/ScanBMA algorithm (without self-directed edges) for DREAM4 time series network 1

Regulator Gene	Target Gene	Predicted Probability
G1	G5	1.000000000000000
G9	G10	1.000000000000000
G5	G1	0.999999999999998
G8	G6	0.999999999999684
G7	G4	0.999999999995794
G7	G3	0.999999999978841
G6	G8	0.999999951242269
G7	G9	0.999999496206814
G9	G7	0.999999288786108
G7	G2	0.999920122310305
G10	G9	0.999918516700184
G10	G4	0.998426261538289
G2	G10	0.980824700000000
G10	G3	0.950097700000000
G1	G2	0.934155100000000
G8	G5	0.932858700000000

NOTE: Shown are pairs whose probability exceeds 90%.

methods have been widely used to fulfill this task. A drawback of unsupervised methods is the low accuracy they achieve due to the lack of training data. Presented here are some semi-supervised methods for GRN inference [291]. These methods infer GRNs by predicting the links, i.e., interactions of transcription factors and target genes. These methods employ two machine learning algorithms, support vector machines (SVMs) and random forests (RF) to perform link prediction. These methods make use of unlabeled data for training. Inductive and transductive learning approaches were investigated. Each of these methods adopts an iterative procedure to obtain reliable negative training data from the unlabeled data. Once training data are gathered, semi-supervised link prediction methods are applied to the gene regulatory networks of *Escherichia coli* and *Saccharomyces cerevisiae*. Experimental results show that the transductive learning approach outperformed the inductive learning approach for both organisms. However, no conclusive difference was identified in the performances of SVM and RF. The

TABLE 6.11: Gene regulation pairs predicted by TimeDelay-ARACNE algorithm for DREAM4 time series network 1

Regulator Gene	Target Gene
G1	G2
G1	G4
G1	G5
G1	G6
G1	G8
G2	G3
G2	G9
G3	G8
G3	G10
G5	G2
G5	G7
G6	G2
G6	G3
G7	G3
G7	G9
G7	G10
G8	G10
G9	G10

results also show that the semi-supervised methods performed better than supervised methods for both organisms.

Using gene expression profiles to infer gene regulatory networks (GRNs) is a key approach to revealing the relationships between transcription factors (TFs) and target genes that may help to uncover underlying biochemical pathways governed by the TFs. Analyzing individual TF and gene associations to a level of biological significance through wet lab experiments is a challenging, costly and time-consuming task [294]. It is therefore useful to adopt computational methods to obtain similar information because results obtained using such methods can be reproduced through less expensive procedures while allowing multiple computational methods to explore data to validate outcomes [138, 235, 244].

Various computational methods for performing unsupervised, supervised and semi-supervised inference of GRNs are available. These methods employ a variety of techniques ranging from Boolean [214] and Bayesian networks [5, 374] to compressive sensing [73]. Integrated toolkits combining different network inference methods are available [180].

Most of these methods are unsupervised. In previous studies, several authors have shown that supervised and semi-supervised methods outperformed unsupervised methods [244]. Supervised methods, unfortunately, require training data containing both positive and negative examples, which may be difficult to gather. In contrast to supervised methods, semi-supervised methods can work with a large number of unlabeled examples, which are easier to obtain [244]. Presented here are semi-supervised methods capable of predicting TF-gene interactions in the presence of unlabeled training examples [291].

Figure 6.12 illustrates a gene regulatory network between the *E. coli* transcription factor (TF) named FNR and several target genes. The network was created using true TF-gene interactions obtained from RegulonDB [316]. In general, if there exists an experimentally verified interaction between a TF and a target gene, the interaction is considered known. The known interactions are generated through wet lab or sometimes dry lab experiments indirectly associated with wet lab experiments and manually curated based on experimental outputs. Conversely, TF-gene interactions that are not yet experimentally verified are considered to be unknown. In Figure 6.12, solid links (or edges) represent known interactions and dashed links represent unknown interactions. Known interactions comprise positive examples and unknown interactions serve as unlabeled examples.

6.2.2.1 Network Inference and Link Prediction

Computational methods for inferring gene regulatory networks (GRNs) can be categorized as supervised, semi-supervised, and unsupervised [244]. Supervised and semi-supervised inference methods involve the training of a mathematical model that serves as a classifier of an unknown entity. Training samples are gathered to be used as inputs in a classifier training process. By comparison, an unsupervised inference method involves neither training nor training samples.

Supervised and semi-supervised inference methods differ primarily based on whether training samples are labeled. In a supervised inference method, each training sample is labeled positive or negative. In a semi-supervised inference method, some training samples are labeled while others are unlabeled. The training samples for the supervised and semi-supervised inference methods are obtained from known and unknown TF-gene interactions.

In the supervised methods, the training sample set contains both positive and negative samples. A known interaction is used as a positive sample. Obtaining a negative sample, on the other hand, is more challenging due to the lack of biological evidence to indicate that there is no

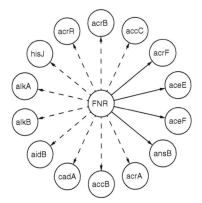

FIGURE 6.12: Diagram showing true regulatory relationships between *E. coli* transcription factor FNR and several target genes. Solid links (or edges) represent known interactions and dashed links represent unknown interactions. Known interactions comprise positive examples and unknown interactions serve as unlabeled examples.

regulatory connection between a transcription factor and a target gene [138].

One training approach for a supervised inference method is to assume that an unknown interaction is a valid negative sample [138, 266]. However, since such an unknown interaction has not been verified experimentally, and some of them may turn out to be positive samples, this training approach is risky. As an example, in the first release of RegulonDB (1.0), 533 regulatory interactions were identified. This number was increased to 4,268 in the later release (8.6), meaning that at the time of the first release 3,735 interactions that were unknown later turned out to be actually valid interactions [178, 316].

Given this possible scenario, a semi-supervised inference method is more suitable. In the training process of the semi-supervised method, an unknown interaction is treated as an unlabeled sample. Therefore, a semi-supervised inference method utilizes positive and unlabeled samples in training to predict gene regulatory networks (GRNs).

Semi-supervised methods infer a GRN by predicting links, i.e., interactions between transcription factors and target gene, in a GRN. Four transcription factors were selected from *E. coli*, namely ARCA, CRP, FIS and FNR. Similarly four transcription factors were selected from *S. cerevisiae*, namely REB1, ABF1, CBF1 and GCN4. The four specific transcription factors (TFs) were chosen because they have the largest numbers, ranging from 100 to 400, of known interactions with target genes in the respective organisms. These known interactions are used as positive examples in GRN experiments. By utilizing an iterative procedure, unlabeled examples (i.e., unknown interactions) are refined to get more reliable negative examples for all four TFs for both organisms.

Inductive and transductive learning approaches are adopted for GRN inference through link prediction using semi-supervised methods. With the inductive learning approach, a model is learned from a training dataset, and then applied to a separate testing dataset that is disjoint from the training dataset; hence any information concerning the test data is not seen or used during creation of the model [262]. On the contrary, transductive learning builds a model based on both the training data and information from the testing dataset, and the model is then applied to the test data [371]. Two machine learning algorithms were employed, namely support vector machines (SVMs) and random forests (RF), in this GRN inference study. The prediction accuracies of both algorithms for the chosen transcription factors of *E. coli* and *S. cerevisiae* are calculated and compared.

A comprehensive assessment of network inference methods on both *E. coli* and *S. cerevisiae* was conducted [247, 248] using two *E. coli* regulatory databases: EcoCyc (*http://ecocyc.org/*) [200] and RegulonDB (*http://regulondb.ccg.unam.mx/*) [316]. RegulonDB was used here because it is a popular database for benchmark experiments. The latest version (8.6) of RegulonDB contains 4,268 known TF-gene interactions obtained from the *E. coli* K-12 strand [316]. Hence, the gene expression datasets specifically generated from *E. coli* K-12 were used. These datasets had GEO accession numbers GSE21869 [25], GSE10158 [221], GSE12411 [11], GSE33147 [121], and GSE17505 [151]. All datasets are freely available at Gene Expression Omnibus (GEO) (*http://www.ncbi.nlm.nih.gov/geo/*) and were produced with the Affymetrix *E. coli* Antisense Genome Array that contains 7,312 probe sets for *E. coli* gene expression analysis.

Three well established *S. cerevisiae* regulatory databases were recently evaluated [246, 247, 249]. These databases include YEASTRACT (*http://www.yeastract.com/*) [4], a map of conserved regulatory sites (*http://fraenkel.mit.edu/improved_map/*) by Fraenkel, et al., [242], and the database described in [173]. Fraenkel's database contains high quality TF-gene interactions used as positive examples in the study presented here.

Five gene expression datasets were selected for *S. cerevisiae*. These datasets have GEO accession numbers GSE30052 [78], GSE12221 [328], GSE12222 [328], GSE40817 [410], and GSE8799 [281]. All the yeast datasets were created using Affymetrix Yeast Genome 2.0 Array containing 5,744 probe sets for *S. cerevisiae* gene expression analysis.

TABLE 6.12: The number of positive and unlabeled examples for each transcription factor of *E. coli* and *S. cerevisiae*, respectively

E. coli			S. cerevisiae		
TF	Positive	Unlabeled	TF	Positive	Unlabeled
CRP	390	770	REB1	217	1,776
FNR	239	921	ABF1	199	1,794
FIS	200	960	CBF1	164	1,829
ARCA	139	1,021	GCN4	120	1,873

Expression vectors were extracted from TFs and genes present in the RegulonDB and Fraenkel's databases, respectively. An expression matrix was created containing the expression vectors for each of the gene expression datasets mentioned above. The *E. coli* expression matrices contained 1,161 gene expressions vectors and the *S. cerevisiae* expression matrices contained 1,994 gene expressions vectors. These matrices were scaled to zero mean and unit standard deviation.

Positive examples were created using the known interactions found in RegulonDB and Fraenkel's database for *E. coli* and *S. cerevisiae*. To obtain unknown interactions, all possible combinations of available TF and gene pairs were generated. Each of these combinations was considered as an unknown interaction provided that it did not exist in RegulonDB or Fraenkel's database for *E. coli* or *S. cerevisiae*. These unknown interactions were treated as unlabeled examples and all the interactions were separated based on the TFs. For each organism, the four TFs that had the largest number of interactions were chosen and used to perform the experiments presented here.

Table 6.12 lists the number of positive and unlabeled examples for each chosen TF. The Positive columns in the table show the total number of known interactions for each TF in *E. coli* and *S. cerevisiae*.

Both supervised and semi-supervised methods work under the principle that if a gene is known to interact with another gene, any other two genes containing similar gene expression profiles are also likely to interact [266]. Based on this principle, feature vectors for TFs and genes are constructed by concatenation of their expression profiles. Hence, the resulting feature vectors contain twice as many features as the original individual expression vectors. While concatenating two profiles, orders were considered, which means, if gene G_1 was known to regulate gene G_2, the feature vector $V(G_1, G_2)$ could only be created such that expression values of gene G_1 were followed by expression values of gene G_2. In other words, the feature vector $V(G_1, G_2)$ implies that gene G_1

1: $P, V \leftarrow Positive$;
2: Initialize(N_0, $Unlabeled$);
3: Create(T_0, N_0, $Unlabeled$);
4: $K \leftarrow Loop\_Count$;
5: $k \leftarrow 0$;
6: **while** $k \neq K$ **do**
7: Train($Model_k, P, N_k$);
8: Validate($Model_k, V$);
9: Classify($Model_k, T_k$);
10: $N_{k+1} \leftarrow$ Extract(T_k);
11: Create($T_{k+1}, N_{k+1}, Unlabeled$);
12: $k \leftarrow k + 1$;
13: **end while**

FIGURE 6.13: Algorithm of semi-supervised link prediction methods.

regulates gene G_2 but the opposite statement may not necessarily be true. After concatenation, the resulting feature vectors were scaled to zero mean and unit standard deviation.

6.2.2.2 Semi-supervised Methods

Two machine learning algorithms, namely support vector machines (SVM) and random forests (RF), were employed for link prediction. SVM analysis was done using the LIBSVM package in R [72]. RF analysis was performed using the randomForest package in R [233].

Figure 6.13 presents the algorithm of the semi-supervised methods for link prediction. Initially, positive and unlabeled examples were present, where positive examples represented known TF-gene interactions and unlabeled examples represented unknown interactions. These examples were converted to feature vectors using the method described earlier in this section, where the feature vectors were used as inputs of the SVM and RF algorithms as appropriate. An iterative procedure was executed whereby a classification model was obtained and validated in each iteration.

In Figure 6.13, *Positive* represents the set of all positive examples obtained from known interactions of each individual transcription factor of an organism, *Unlabeled* indicates the set of all unlabeled examples for the organism, P the positive training dataset, N_k the negative training dataset during iteration k, V the validation dataset, and T_k the testing dataset during iteration k. The validation set V contains only positive examples since true negative examples are not available. All the predictions were made on the examples in V, and prediction accuracies of inference methods were calculated by comparing true labels with predicted labels. The *Positive* set is evenly divided into two disjoint subsets, P and V and each subset contains approximately the same number of positive examples. For a given transcription factor (TF), P and V remain the same throughout all the iterations for both inductive and transductive learning approaches. T_k comprises only unlabeled examples, which are used in the iterative procedure to produce reliable negative examples.

In Figure 6.13, note iteration k, $0 \leq k \leq K$. Referring to step 7, Train$(Model_k, P, N_k)$, the SVM or RF algorithm is trained using P and N_k. A binary classification model denoted $Model_k$ is obtained. Referring to step 8, Validate$(Model_k, V)$, this (SVM or RF) model, $Model_k$, is then applied to the validation dataset V to predict the labels of the examples in V. The prediction accuracy of the model is calculated by dividing the number of correctly predicted examples in V by the total number of examples in V. In addition, $Model_k$ is applied to the testing set T_k to classify the unlabeled examples as shown in step 9, Classify$(Model_k, T_k)$. Both SVM and RF algorithms are able to assign probabilistic weights to their classification results. Using these probability values, reliable negative examples are extracted from the set T_k by choosing the bottom $|P| = |N_k|$ unlabeled examples in T_k that have the lowest probability of being positive. As shown in step 10, these extracted $|P| = |N_k|$ negative examples are collected and stored in N_{k+1}, which, together with P, are used to train the SVM or RF algorithm in iteration $k + 1$. Notice that iteration 0 is a special case, in which $|P| = |N_0|$ examples are randomly selected from the *Unlabeled* set and stored in N_0, as seen in step 2, Initialize$(N_0, Unlabeled)$. Notice also that this is a balanced binary classification, since $|P| = |N_k|$ throughout all the iterations.

The difference between the inductive and transductive learning approaches is whether the sets N_k and T_k are disjoint. For the inductive learning approach, T_k is created such that N_k and T_k are disjoint. For the transductive learning approach, however, T_k includes all the available unlabeled examples (i.e., all the examples in the *Unlabeled* set). More precisely, in iteration k, for the inductive learning approach, $T_k = Unlabeled - N_k$ and $N_k \cap T_k = \varnothing$; for the transductive learning approach, $T_k = Unlabeled$ and $N_k \subseteq T_k$ (as seen in step 3, Create$(T_0, N_0, Unlabeled)$), and step 11, Create$(T_{k+1}, N_{k+1}, Unlabeled)$, in Figure 6.13). The iterative procedure is executed for each of the four chosen transcription factors (TFs) in each organism. Table 6.13 lists the number of examples in P, N_k, T_k and V for each TF in *E. coli* and Table 6.14 shows the data for each TF in *S. cerevisiae*. The Total column in each table shows the number of training examples, which is equal to $|P| + |N_k|$, used in each iteration. The Inductive and Transductive columns in these

TABLE 6.13: Number of examples in P, N_k, T_k and V for each TF in *E. coli*

TF	P	N_k	Total	T_k Inductive	T_k Transductive	V
CRP	195	195	390	575	770	195
FNR	120	120	240	801	921	119
FIS	100	100	200	860	960	100
ARCA	70	70	140	951	1,021	69

TABLE 6.14: Number of examples in P, N_k, T_k and V for each TF in *S. cerevisiae*

TF	P	N_k	Total	T_k Inductive	T_k Transductive	V
REB1	108	108	216	1,668	1,776	109
ABF1	99	99	198	1,695	1,794	100
CBF1	82	82	164	1,747	1,829	82
GCN4	60	60	120	1,813	1,873	60

tables show the number of unlabeled examples in the testing set T_k used by the inductive and transductive learning approaches.

In all the experiments, the radial basis function (RBF) kernel was used for support vector machines (SVMs) with all other parameters set to default values. With random forests (RF), all parameters were set to default values and the number of trees was 500. For statistical consistency and fair comparisons, the *Positive* and *Unlabeled* sets were kept the same for both SVM and RF in the initial iteration (i.e., iteration 0) for each chosen TF.

Semi-supervised methods: comparative experiments

A series of experiments shows the performance results of semi-supervised methods for the datasets described earlier in this chapter. The performance of each method was measured by its prediction accuracy defined as the number of correctly predicted examples in a validation dataset divided by the total number of examples in the dataset (cf. Figure 6.13). Figures 6.14(a) and 6.14(b) compare the transductive and inductive learning approaches with the SVM and RF algorithms using the *E. coli* transcription factor ARCA and dataset GSE21869. Figures 6.15(a) and 6.15(b) compare the transductive and inductive learning approaches with the SVM and RF algorithms using the *S. cerevisiae* transcription factor REB1 and dataset GSE12222.

As seen in Figures 6.14(a) and 6.15(a) for SVM, the transductive learning approach yields higher prediction accuracies than the inductive learning approach. The performance of the transductive learning approach becomes stable as the number of iterations in the semi-supervised methods increases. On the other hand, the performance of the inductive learning approach tends to fluctuate up and down with high frequency. RF exhibits a similar pattern with respect to the relative performance of the transductive and inductive learning approaches.

It is worth pointing out that the accuracies of the semi-supervised methods are relatively low when only one iteration (i.e., iteration 0) is executed (cf. Figure 6.13). During iteration 0, the negative training set N_0 is comprised of randomly selected unlabeled examples. On the other hand, starting from the second iteration (i.e., iteration 1), the semi-supervised methods pick unlabeled examples having the lowest probabilities of being positive and use them as negative training examples. These unlabeled examples with the lowest probability form more reliable negative training data than the randomly selected unlabeled examples, hence yielding higher accuracies.

In other experiments, the transductive learning approach was utilized with the number of iterations fixed at 15. Semi-supervised link prediction methods with both SVM and RF were applied to the different gene expression datasets for transcription factors selected from *E. coli* and *S. cerevisiae*.

Experimental results for *E. coli* are summarized in Figures 6.16 and 6.17. SVM yield higher prediction accuracies than RF on the datasets GSE33147 and GSE17505 for the transcription factor ARCA while RF performed better than SVM on the dataset GSE10158. The two machine learning algorithms exhibited similar performance on the datasets GSE12411 and GSE21869. For the transcription factor CRP, SVM and RF exhibited similar performance on all datasets except GSE17505, where SVM was more accurate than RF. Significant discrepancies were observed in the prediction accuracies with the transcription factor FIS, where RF outperformed SVM on all datasets except GSE33147. Finally, SVM and RF did not show any major differences in GRN prediction for the *E. coli* transcription factor FNR.

On the other hand, the performances of SVM and RF were nearly identical across all the datasets for all the *S. cerevisiae* transcription factors studied (cf. Figures 6.18 and 6.19). There are only two instances where significant differences were observed. The first instance is on the dataset GSE8799 for the *S. cerevisiae* transcription factor ABF1, as seen in Figure 6.18(b). The second is on the dataset GSE12222 for the

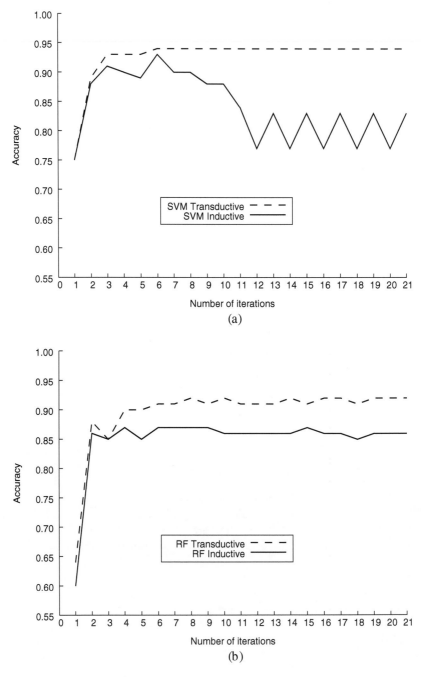

FIGURE 6.14: Performance comparison of the transductive and inductive learning approaches based on the *E. coli* transcription factor ARCA and dataset GSE21869 with the SVM algorithm and the RF algorithm.

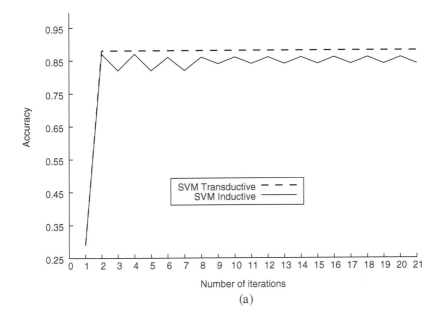

(a)

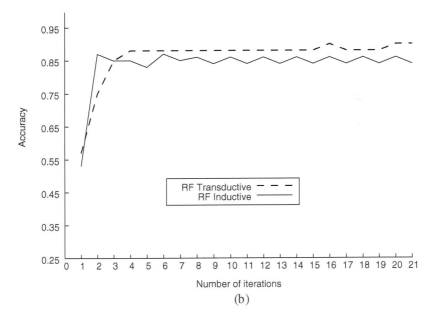

(b)

FIGURE 6.15: Performance comparison of the transductive and inductive learning approaches based on the *S. cerevisiae* transcription factor REB1 and dataset GSE12222 with the SVM algorithm and the RF algorithm.

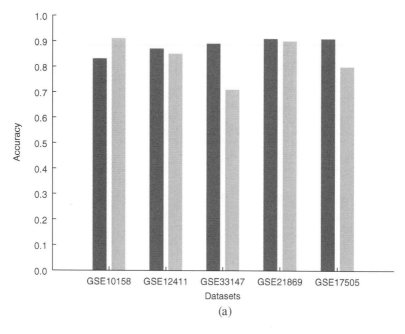

(a)

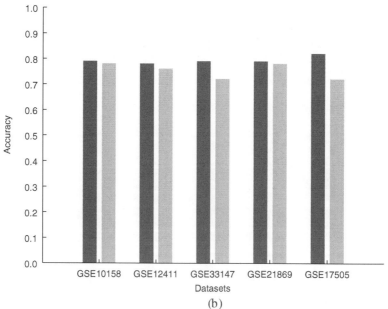

(b)

FIGURE 6.16: Performance comparison of the SVM and RF algorithms with the transductive learning approach on five gene expression datasets (GSE10158, GSE12411, GSE33147, GSE21869 and GSE17505) and two transcription factors of *E. coli* including (a) ARCA and (b) CRP. Dark bars represent SVM and light bars represent RF.

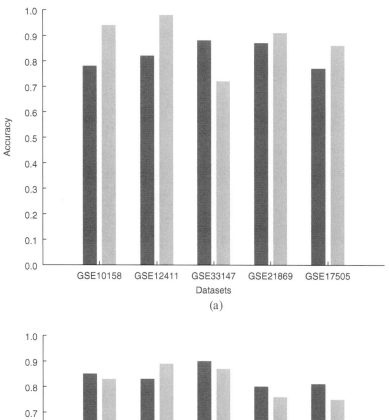

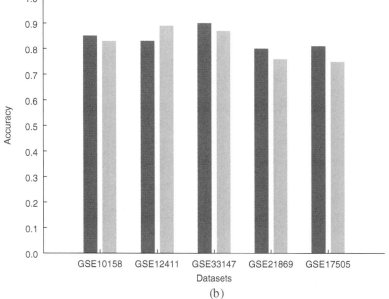

FIGURE 6.17: Performance comparison of the SVM and RF algorithms with the transductive learning approach on five gene expression datasets (GSE10158, GSE12411, GSE33147, GSE21869 and GSE17505) and two other transcription factors of *E. coli* including (a) FIS and (b) FNR. Dark bars represent SVM and light bars represent RF.

S. cerevisiae transcription factor GCN4, as seen in Figure 6.19(b). In both instances, RF outperformed SVM.

Semi-supervised methods: experimental observations

For *E. coli* and *S. cerevisiae*, four transcription factors with significant numbers of known TF-gene interactions were chosen for each organism to evaluate the effectiveness of the semi-supervised methods. These methods employ an iterative procedure with a transductive or inductive learning strategy to obtain more reliable negative training examples. Experimental results indicate that the transductive learning approach consistently outperformed the inductive learning approach on the datasets tested. The results also show that after certain iterations, the prediction accuracy of the transductive learning approach tends to converge. For many experiments, the convergent point was identified within the first 10 iterations. To provide more reliable and consistent findings, the number of iterations was fixed at 15 in the remainder of the experiments. With 15 iterations, no significant difference in prediction accuracies was observed between the SVM and RF algorithms using the transductive learning approach.

On the other hand, the inductive learning approach did not exhibit a clear convergent point; instead, the inductive learning approach exhibited waving patterns (cf. Figure 6.14). Such a behavior might arise from the fact that in the inductive learning approach a portion of TF-gene interactions is not used. As explained previously, in iteration k, $T_k = Unlabeled - N_k$, and some unlabeled examples are extracted from T_k to get negative training examples in N_{k+1} that are used in iteration $k + 1$. T_k does not contain the TF-gene interactions in N_k. Hence, when the TF-gene interactions in T_k are ranked in the descending order of their probabilities of being positive, they do not contain the interactions in N_k, and therefore will not be in N_{k+1}. It is likely that N_k may contain some TF-gene interactions that are very reliable negative examples. As a consequence, these reliable negative examples will not be in N_{k+1} used for training the machine learning algorithms (SVM or RF) in iteration $k + 1$. Hence, when those examples are in the training set, the trained model yields a high accuracy; otherwise, the trained model yields a low accuracy. This explains why the inductive learning approach exhibits waving patterns of prediction accuracies. Note that in the transductive learning approach, $T_k = Unlabeled$ and those very reliable negative examples are considered in every iteration. As a consequence, the performance of the transductive learning approach becomes stable when sufficient high-quality negative examples are collected after a certain number of iterations (e.g., 15 iterations).

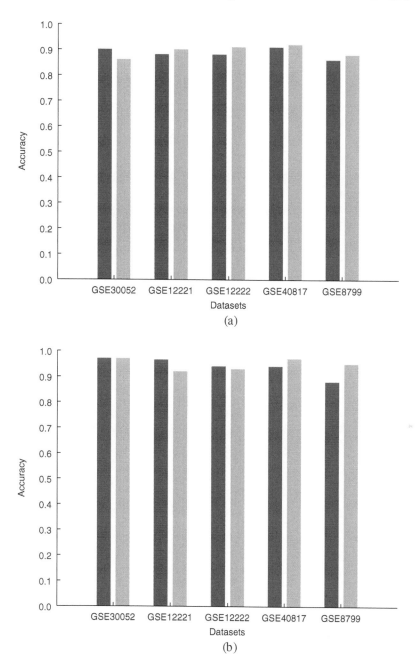

(a)

(b)

FIGURE 6.18: Performance comparison of the SVM and RF algorithms with the transductive learning approach on five gene expression datasets (GSE30052, GSE12221, GSE12222, GSE40817 and GSE8799) and two transcription factors of *S. cerevisiae* including (a) REB1 and (b) ABF1. Dark bars represent SVM and light bars represent RF.

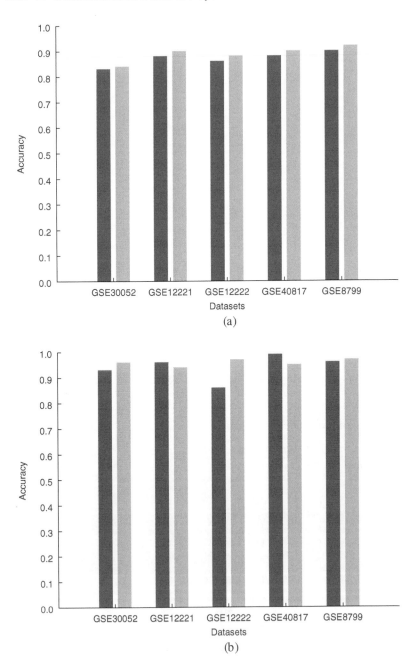

(a)

(b)

FIGURE 6.19: Performance comparison of the SVM and RF algorithms with the transductive learning approach on five gene expression datasets (GSE30052, GSE12221, GSE12222, GSE40817 and GSE8799) and two other transcription factors of *S. cerevisiae* including (a) CBF1 and (b) GCN4. Dark bars represent SVM and light bars represent RF.

The experimental results were obtained using the radial basis function (RBF) kernel for the SVM algorithm and 500 trees for the RF algorithm. The linear kernel and polynomial kernel available in the LIBSVM package were also tested using default parameter values on all gene expression datasets and for all transcription factors. For the RF algorithm, tests were conducted with 100, 500 and 1,000 trees. The number of iterations used in the semi-supervised methods was fixed at 15. The results were similar to those presented above and indicate that all three kernels, namely RBF, linear and polynomial kernels, work well. Furthermore, the number of decision trees used in the RF algorithm had little impact on its performance provided the number was sufficiently large (e.g., at least 100). However, with too many trees (e.g., more than 1,000), the time of the RF algorithm increased substantially due to the time required to build the trees.

The performance of the SVM algorithm clearly converges with the transductive learning strategy in both organisms in that after a certain number of iterations (e.g., 15 iterations), the accuracies of the algorithm do not vary much (cf. Figures 6.14(a) and 6.15(a)). The reason is that after 15 iterations, the algorithm has identified the most reliable negative training examples, which remain the same in subsequent iterations. Therefore, the models created for subsequent iterations by the negative training examples and positive training set P remain almost the same, and hence always make the same predictions.

On the other hand, Figures 6.14(b) and 6.15(b), illustrate that with the transductive learning strategy, there are slight variations in the performance of the RF algorithm even after 15 iterations. Although those variations seem negligible, and lead to qualitative conclusions, a close look at the machine learning algorithms explains why the variations occur. In general, the SVM algorithm systematically attempts to find a hyperplane that maximizes the distance to the nearest training example of any class [186]. There is no randomness associated with the algorithm. In contrast, the RF algorithm randomly picks training examples and features in the training examples to build decision trees [50]. Due to the randomness associated with the RF algorithm, the strong convergence was not observed.

It is worth pointing out that the semi-supervised methods with the transduction learning strategy performed better than supervised methods. Supervised methods treat randomly selected unlabeled examples as negative examples and use them with positive examples in P to train the machine learning algorithms [138, 266]. As shown in Figures 6.14 and 6.15, executing the first iteration without iteratively refining the unlabeled examples to obtain more reliable negative examples (as done in the supervised methods) performs worse than executing several (e.g.,

15) iterations (as done in the semi-supervised methods). This suggests that the semi-supervised methods may be better than the supervised methods.

Semi-supervised methods: experimental findings

Semi-supervised methods for link prediction infer gene regulatory networks using both inductive and transductive learning approaches. To utilize available unlabeled examples and effectively extract reliable negative examples, support vector machines (SVM) and random forests (RF) were used. Both SVM and RF assign probabilistic weights to their classification results. Negative examples that have the lowest probability of being positive are picked from the testing set. Experimental results show that the negative examples chosen this way yield better performance than negative examples that were randomly selected from unlabeled data.

In addition, experimental results show that the transductive learning approach outperformed the inductive learning approach and exhibited a relatively stable behavior for both SVM and RF algorithms on the datasets used in the study. Algorithmic parameters such as different kernels for SVM and different numbers of decision trees for RF did not yield significantly different results. Furthermore, there was no clear difference in the performance of SVM and RF for either the prokaryotic organism (*E. coli*) or the eukaryotic organism (*S. cerevisiae*). The experimental results also show that the semi-supervised methods with the transductive learning strategy were more accurate than supervised methods [138, 266] since the semi-supervised methods adopt an efficient iterative procedure to get more reliable negative training examples than those used by the supervised methods.

A shortcoming of semi-supervised methods is that sometimes insufficiently known TF-gene interactions are available for certain transcription factors or there exist no known TF-gene interactions for some organisms. Under this circumstance, semi-supervised methods may yield low prediction accuracies due to the lack of reliable training data. One possible approach for dealing with organisms with only unknown TF-gene interactions or unlabeled examples is to use SVM and RF algorithms to assign probabilistic weights to their classification results. Choose positive and negative examples from the testing set that have the highest and lowest, respectively, probability of being positive, then use the more reliable positive and negative training data to obtain a potentially improved classification model.

More investigative research into the performance of semi-supervised methods will likely lead to favorable comparisons against other machine learning algorithms.

6.2.3   Miscellaneous Inference Methods

Gene regulatory network (GRN) inference methods presented in this chapter are relatively well established. Each method has advantages and disadvantages. In any new technology, such as computational biology, novel approaches are developed by combining existing approaches in a best-of-breed manner to capture the advantages of well established methods.

Solving the GRN inference problem using a purely ordinary differential equation (ODE) approach is not easily scalable to the larger genomes of many species. However, by combining ODE with other computational methods such as particle swarm optimization (PSO) [198], an efficient heuristic solution to the GRN inference problem may be possible. Originally intended for studying social behavior, PSO relates swarm behaviors of organisms such as flocks of birds or schools of fish. PSO has been applied to various unrelated problem solving techniques for efficiently identifying solutions through iterative improvement [224].

The importance of more sophisticated statistical and mathematical modeling for GRN inference is well understood. When a gene transcript is present in a gene expression dataset, accurately modeling the biological system that predicts the probability of translating that transcript into protein is a sub-problem of the larger GRN inference problem. Nevertheless, modeling these transcription-translation networks is vital and sufficiently complex as to require additional hybrid approaches to regulatory modeling [278].

CHAPTER 7

Cloud-Based Biological Data Processing

CONTENTS

7.1 INTRODUCTION

Advances in computing power occur in parallel with advances in biological knowledge. The recent sequencing [86] of the complete genome of the sea urchin (*Strongylocentrotus purpuratus*), for instance, revealed the enormous complexity of its immune system. Further analysis revealed that gene clusters associated with this uniquely complex immune system have been present in animal genomes for longer than previously known. These findings have important implications that make the sea urchin a valuable model for further genomic research. We can expect the explosive growth in biological knowledge to continue as researchers leverage ever-advancing technologies like cloud computing.

Cloud computing is a model for enabling ubiquitous, convenient, on-demand network access to a shared pool of configurable computing resources (e.g., networks, servers, storage, applications, and services) that can be provisioned and released rapidly with minimal management effort or service provider interaction *(see NIST: SP800-145.pdf)*. These characteristics are typical of cloud computing services:

- *On-demand self-service provisioning*
  Once approved for a cloud service, a cloud user typically is enabled to use a graphical user interface (GUI) or command-line interface (CLI) with options that vary depending on user approval level. The recent development of the virtualized PC has been one on the major contributing factors behind the popularity of cloud computing. A virtual machine can be provisioned in a matter of minutes with specific CPU, memory and disk storage resources as required by the application. This provisioning is accomplished by the point and click of a mouse or the running of a command line script without human intervention.

- *High-speed network access*
  Ubiquitous high-speed internet access has brought the vast resources owned by cloud service providers to any entity or individual that can run a browser on any large or small device. There is no need for a "thick-client" with powerful processors, massive memory and disk storage. A cloud services user with a "thin-client" is just as enabled as the cloud services user with a "thick-client".

- *Resource pooling*
  Large cloud service providers such as Amazon, Google and Microsoft, leverage economies of scale to purchase very large quantities of commodity hardware at a relatively small unit rate. By using software designed and developed for a cloud service environment, these vast hardware resources are shared in a secure multi-tenant environment.

- *Rapid elasticity*
  One of the primary objectives for a cloud user is to "pay as you go." There are cycles in most business operations whereby at certain times for predictable reasons, business picks up. At other times, business is slow. A business that can't handle peak periods, will lose customers to business that can handle the peak periods. Rather than invest in expensive computing equipment to handle the peak periods and have that equipment sit idle during slow periods, it makes good business sense to pay for computing resources only when needed. This is an attractive feature of cloud computing.

- *Measured service*
  Cloud computing turns information technology (IT) into a utility, much like electricity. The business model for electrical service is clear. A customer uses electricity and is periodically billed for the usage in concise understandable terms. Cloud computing converts IT spending from capital expenses (CapEx) to operational

expenses (OpEx). Prior to cloud computing as a viable strategic business option, the expansion of computing resources was a CapEx requiring computing equipment and brick-and-mortar construction in a new or existing facility. Cloud computing offers business a strategic choice, depending on inter-related business factors, between buying or renting computing resources.

Cloud computing service models generally fall into one of the following categories *(see NIST: SP800-145.pdf)*:

- *Private (local) cloud service*
  A private or local cloud service is dedicated to one organization and may be administered by employees of the organization or by a different organization. This cloud service may be located on premises or off premises at a remote location. While we generally conceive of a cloud service as always located at a remote site, that is not always the case. Customers of a private or local cloud service often are business units within the organization.

- *Public cloud service*
  A public cloud service is available to the general public. Well-known public cloud services are Amazon Web Services (AWS), Google's AppEngine and Microsoft's Azure. Large public cloud service providers manage very large data centers around the globe. Multiple data centers offer customers the option to maintain multiple copies or replicas of their data at geographically diverse locations for disaster recovery (DR) protection. Such data replication helps ensure high levels of data availability.

- *Community cloud service*
  A community cloud service is shared by multiple organizations with common interests. In the United States, institutions of higher education within the same state pool resources and talent to establish community cloud service in their state. Many aspects of the business of higher education are similar throughout all institutions. Peak periods of activity in higher education are fall and spring semester registration. State government regulations are sometimes common challenges for many higher education institutions in the same state. A community cloud service offers high quality IT services to students and researchers within rigid budgetary constraints.

- *Hybrid cloud service*
  A hybrid cloud service combines features of one or more cloud computing service models described above.

Cloud computing deployment models include the following:

- *Infrastructure as a service (IaaS)*
  With an Infrastructure as a Service (IaaS) cloud computing deployment model, the cloud service customer has the greatest flexibility and the greatest responsibility regarding the computing resources that are available. The customer is granted full system administration access to a dedicated pool of server, memory and disk storage resources by the IaaS cloud service provider. Within this dedicated pool of resources, the customer self-provisions and self-supports any number of virtual machines (VMs). While the operating system is pre-installed in each VM, the IaaS cloud service customer is responsible for installing and patching any additional software needed for the intended applications. Amazon Web Services (AWS) is among the largest IaaS cloud service providers in the IaaS cloud service industry.

- *Software as a service (SaaS)*
  With a Software as a Service (SaaS) cloud computing deployment model, the customer generally has very little if any system administration privileges. The customer is granted full application administrator access to an application preinstalled on a VM at the SaaS cloud service provider's premises. Within this application, the customer manages the application configuration and/or user access to the application. Application patching and other routine application maintenance are provided by the SaaS cloud service provider on an agreed upon schedule and are not the responsibility of the customer. Enterprise Resource Planning (ERP) administrative software is a common application offered by SaaS cloud service providers.

- *Platform as a service (PaaS)*
  With a Platform as a Service (PaaS) cloud computing deployment model, the customer is granted limited system administration access to a virtual server provisioned with specific resources at the provider's premises. Within this virtual server, the customer installs and manages one or more applications. Operating system patching and other routine system maintenance are provided on an agreed upon schedule and is not the responsibility of the customer. Microsoft Azure is a popular PaaS cloud service in the industry.

7.1.1   Motivation

Organizations evaluate risk tolerance before deciding on sizing and purchasing computing hardware to meet service demand. By under-provisioning computing resources, some customers may not be served, so there is a risk of losing a customer and reducing profits. In addition, how does a company factor its reputation into the risk equation? Overprovisioning presents a risk that resources will remain idle. Support costs to run unneeded resources are wasted, affecting corporate profits.

An entrepreneur wishing to start up a Web application company using cloud computing services, as an example, need not worry about over-purchasing or under-purchasing computing equipment to launch a new service. Cloud-based resources can be readily increased or decreased as demand dictates. A temporary surge in demand may require dramatically increasing the number of servers 10-fold or 100-fold in a matter of days. After a surge subsides and customer business activity levels off to a steady state well below the surge, cloud-based server resources should be as sharply cut back. Elasticity of cloud services provides the customer the ability to "pay as you go" and not pay for resources that are not needed.

Computing resource elasticity is equally valuable to established companies and startups. Cloud computing services are applicable when demand for a service varies over time or demand is unknown. One of the busiest periods for retailers each year is "Black Friday" in November. The elasticity of cloud computing services could very well determine whether an online web-site crashes or rides smoothly through such busy peak periods.

With parallel batch processing on cloud computing, cost is the virtually the same to run a job for 1,000 hours on one server as the cost to run the same job in one hour using 1,000 servers [23]. Cloud service providers are more equipped than small companies to handle threats such as distributed denial of service (DDoS) attacks. A growing amount of research focuses on the cloud. The cloud can provide a researcher with significant computing resources without large capital investments. Limited access to computational infrastructure is a serious bottleneck for many researchers. The increasingly rich set of cloud computing services offers a very attractive environment for scientists to perform data analytics.

Cloud computing resources typically consist of commodity equipment running commodity operating systems, like Linux and Windows, that most application developers already understand. The generic nature of cloud computing allows resources to be purchased on demand, especially to augment local resources for specific large or time-critical tasks.

It is important to develop frameworks to conduct data-intensive scientific discovery. The pay-as-you-go cloud computing model provides an option for the computational and storage needs of such computations [146]. The emergence of data-intensive computational fields such as bioinformatics, chemical informatics, weather forecasting and Web mining drives the need to process massive amounts of data using frameworks designed with cloud-based parallel and distributed processing. Pairwise alignment, multiple alignment, distance calculation, clustering, multidimensional scaling and visualization of gene sequences are computationally intensive. Similarly, interactive parallel computations are required to determine differential equation parameters in resolving gene network inferences. The framework used should be straightforward, like MapReduce, and should not require extensive programming skills to master. Additionally, the framework should manage critical administrative functions like fault tolerance, master-worker task scheduling, etc. The researcher should then be able to focus on the solution to the research problem at hand and not on the infrastructure of the framework environment. This is one of the attractions of cloud computing.

7.2   DATA PROCESSING IN THE CLOUD

Biological sequencing technology continually improves as the sequencing cost per nucleotide continually drops. This situation could create a bioinformatics bottleneck if the sequence data cannot be processed and analyzed as quickly as it is produced [21]. Processing and analysis of genomic sequence data include sequence assembly, annotation, database BLAST searches and other tasks. Cloud computing offers some relief to the bottleneck in that biological sequence analysis can be performed in a decentralized mode. By using temporary cloud-based resources, small biotech corporations need not rely on larger firms whose computing facilities have the capacity for the many hours of computation required for thorough sequence analysis. Customized applications dedicated to cloud-based sequence analysis processing, such as Cloud Virtual Resource (CloVR), are available to genomics researchers [20]. Clusters of virtual machines running in parallel on cloud resources may help address the bioinformatics bottleneck problem.

Biological data is being sequenced and recorded at exponential rates [1]. When the human genome was sequenced in 2000, 8 billion finished base pairs were recorded at the International Nucleotide Sequence Database Collaboration (INSDC) [57] comprised of the DNA Data Bank of Japan (DDBJ) [280], the European Molecular Biology Laboratory (EMBL) [350], and GenBank at the US National Center for Biotechnology Information (NCBI) [37]. DDBJ, EMBL and GenBank are the three main databases for finished (i.e., fully analyzed and annotated) sequences. Ten years later in 2010, there were 270 billion finished base pairs recorded at the INSDC, which equates to a doubling of base pairs every 18 months. However, in NCBI's Sequence Read Archive (SRA), which houses raw data from next-generation sequencing, there were 25 trillion base pairs recorded as of 2010. Genomic sequence data is among the categories of big data that require non-traditional data processing approaches.

The world's collective genomic data are unlikely to be stored in one repository due to overwhelming storage requirements. Rather, repositories of genomic data will be maintained at research facilities throughout the world and shared. Data representation and API data exchange standards were developed at the NIH Big Data to Knowledge (BD2K) Center [292]. These standards combined with cloud computing technologies enable genomic big data processing and research.

Apache Spark is an open source cluster computing framework that utilizes a distributed memory abstraction (*http://spark.apache.org*) [414]. In certain situations, Apache Spark can help achieve a performance improvement an order of magnitude above that of Apache Hadoop (*http://hadoop.apache.org*). Two applications that can benefit from this approach are iterative algorithms and interactive data mining. Apache Spark SQL is an Apache Spark module that allows parallel processing of data with relational database and traditional procedural instructions in the same application [24]. Apache Spark SQL utilizes a dataframe API, similar to data frame in R [239] (*https://www.r-project.org*). The dataframe API allows performance optimization and relational and procedural code integration. A genomic oriented enhancement to Apache Spark SQL named GenAp improves performance by an order of magnitude in certain interval join situations [208].

Cloud computing services require high availability of the Internet. Google search has been compared to the "Ma Bell dial tone" in describing Internet reliability. While 99.999% (5 nines) reliability may be difficult to achieve consistently year after year, most agree that in general the Internet is reasonably reliable. Nevertheless, contingency planning is needed in case a cloud service provider experiences an unanticipated service interruption.

Several organizations offer cloud compute cycles. Amazon's Elastic Compute Cloud (EC2) *(http://aws.amazon.com)* has capacity for many thousands of virtual machines (VMs). EC2 maintains five different classes of virtual machines providing levels of CPU, RAM and disk resources with price ranging from $0.10 to $0.80 per hour per virtual machine. Amazon offers preconfigured disk images and launches scripts for initializing clusters. Once initialized, a customer copies data into AWS

S3 (Simple Storage System) and executes his or her jobs as if the cluster was dedicated exclusively for its use. For very large datasets, the time required for the initial data transfer can be substantial and depend on the bandwidth of the cloud provider. Once transferred into the cloud the cloud nodes generally have very high-internode bandwidth. Furthermore, Amazon mirrors portions of Ensembl and GenBank for use within EC2 without additional storage costs, thereby minimizing the time and cost to run large-scale analyses of these data [323].

Data transfer fees are costs of using cloud services. One solution to consider is shipping of an encrypted disk drive. Cloud providers may offer import/export data transfer service. As amounts of information available for research grow at exponential rates, a critical need has arisen for tools to evaluate the available information and leverage its value. Google developed an effective approach called MapReduce to evaluate large amounts of information. Soon after Google's proprietary implementation of MapReduce, Apache implemented MapReduce in the open source domain and named it Hadoop (*http://hadoop.apache.org*).

Table 7.1 shows recent pricing for AWS Elastic Compute Cloud (EC2) virtual machines. EC2 also offers virtual machines that are memory optimized and storage optimized at increased rates. Stormbow demonstrates how economical biological data processing can be using cloud computing [417]. The AWS EC2 application developer chooses a global region in which to provision a virtual machine. Amazon has large data centers located throughout the world to ensure geographic diversity if the customer should require it. AWS data centers to choose from include United States, Australia, Japan, Brazil, Ireland and Singapore. The developer chooses the operating system (OS) of the virtual machine to be provisioned. OS options include 32 and 64-bit versions of Ubuntu Linux, RedHat Linux, Windows and others. The developer is responsible for installing and patching all software modules needed on the virtual machine for the specific application. Once a virtual machine is fully configured, the developer may create an Amazon Machine Image (AMI). The developer then can choose the customized image for future virtual machine provisioning. Additionally, virtual machines with certain customized features can be provisioned in any quantities and in any global region. Provisioning virtual machines in multiple regions provides geographic diversity for additional fault tolerance for critical applications.

The developer must ensure that the data to be processed by the cloud application is located where the cloud application can access it. This can be in one of several places such as on the local disk drive of a virtual machine, at Amazon Elastic Block Storage (EBS), Amazon Simple Storage Service (S3) or other locations. There may be a significant

TABLE 7.1: Amazon web services (AWS) elastic cloud computing (EC2) general purpose instance types and pricing

Virtual Machine Size	CPU Cores	Memory GiB	Storage GB	Cost Per Hour ($)
Micro	1	0.615	EBS only	0.020
Small	1	1.7	1 X 160	0.044
Medium	1	3.75	1 X 4 SSD	0.070
Large	2	7.5	1 X 32 SSD	0.140
X-Large	4	15	2 X 40 SSD	0.280
2X-Large	8	30	2 X 80 SSD	0.560

data communication charge incurred in the course of transferring data into the Amazon cloud. Amazon provides the option to transfer large amounts of data on an external hard disk drive which will then be copied to its cloud storage location as requested by the developer. The developer chooses the most appropriate data setup process based on convenience, cost and other factors.

Amazon provides access to a number of large, popular datasets that are freely available. Table 7.2 shows a small sampling of public data sets available in 2014.

7.2.1 MapReduce Parallel Processing

In many situations, a problem can be solved by breaking it into smaller parts, solving the smaller parts simultaneously in parallel, and then gathering the individual results. The time saved with this parallel processing approach is called speed-up.

Amdahl's law (Equation (7.1)) defining speed-up is used in parallel computing to predict the theoretical maximum speed-up using multiple processors, or to simply get a rough estimate of performance improvement [15, 209].

$$Speed\text{-}up = \frac{1}{(1-P) + \frac{P}{S}} \tag{7.1}$$

In Equation (7.1), P is the proportion of a problem that can be divided into parallel computing, and $1 - P$ is the proportion that cannot be parallelized, i.e., the sequential proportion. The maximum acceleration capability lies in S, meaning that P can be accelerated S times by using S parallel processors (i.e., the slaves on Hadoop). The speed-up of a

TABLE 7.2: Recent Amazon Web services (AWS) public data sets samples

Name	Description
Common Crawl Corpus	Web crawl data composed of over 5 billion Web pages
NASA NEX	Climate projections and satellite images of Earth
Human Microbiome Project	Human Microbiome Project data set
1000 Genomes Project	Project initiated in 2008
Japan Census Data	Data sets including 1995 through 2009
Enron Email Data	Publicly released email data
Denisova Genome	Extinct human genome sequence
Google Books Ngrams	Data set of n-gram corpuses
Sloan Digital Sky Survey	Astronomical survey
Million Song Dataset	Collection of popular music
Human Liver Cohort	Characterizes gene expression

program using multiple processors in parallel computing is limited by the time needed for the sequential fraction of the program. For example, assume that a program needs 20 hours using a single processor core and a particular portion of the program (i.e., $1 - P$) which takes 1 hour (i.e., 5%) to execute cannot be parallelized. The remaining 19 hours (i.e., 95%) of execution time can be parallelized (i.e., P). In this case, regardless of how many processors are devoted to a parallelized execution of this program, the minimum execution time cannot be less than 1 hour. Hence the speed-up is limited to at most 20. The law is concerned with the speed-up achievable by parallelizing proportion P of a computation where the improvement has a speed-up of S.

In Amdahl's law, the sequential portion of a task restricts the speed-up performance. As a result, a linear speed-up rate is difficult to achieve. The sequential part of a computational task creates an inevitable constraint on the speed-up rate. Amdahl's law clearly describes the efficiency and limitation of parallel computing and has been widely adopted.

MapReduce provides an easy-to-use programming model that features fault tolerance, resource scheduling, scalability and data locality-based optimizations. Many scientific computation algorithms that rely on iterative computations can be implemented with a MapReduce computation specified for each iterative step. For instance, k-means clustering is often implemented using an iterative refinement technique in which the algorithm iterates until the difference between cluster centers in subsequent iterations, i.e., the error, falls below a predetermined threshold. Each iteration performs two main steps: the cluster assignment and the centroids update step. In the MapReduce implementation, the assignment step is performed in the map task while the update step is performed in the reduce task. Centroid data is broadcast at the beginning of each iteration. Intermediate data communication is relatively costly in k-means clustering as each Map Task outputs data equivalent to the size of the centroids in each iteration. [146]. Pairwise sequence alignment can be parallelized [238]. Additionally, network inference refinement methods are applied [224] where map and reduce phases are iterated until parameters of a set of differential equations match the curve of a known gene expression pattern.

One promising choice is to utilize cloud computing with the MapReduce programming model for a relatively easy-to-implement framework with fault-tolerant capabilities. This framework has been used to successfully solve many large-scale scientific computing problems, especially complex life sciences challenges. The goal of MapReduce is to deploy a large number of time- and memory-consuming tasks to all computing nodes that process tasks in parallel by user-defined algorithms. The flow of the MapReduce process involves one master and multiple slave machines.

MapReduce [96] is a software framework developed and used by Google to support parallel distributed execution of its data intensive applications. MapReduce was developed to address the problem of information explosion. The approach uses a "divide-and-conquer" principle to speed up the processing of large amounts of information. Using MapReduce, a data processing solution consists of map and reduce steps. Google uses this framework internally to execute thousands of MapReduce applications per day, processing petabytes of data, all on commodity hardware. Unlike other parallel computing frameworks that require application developers to explicitly manage inter-process communication, computation in MapReduce is divided into two major phases called map and reduce. Running between the map and reduce phases is an internal shuffle phase for handling intermediate results. The MapReduce framework automatically executes those functions in parallel, and with effective fault tolerance mechanisms for resiliency, over any number (in theory) of virtual processors.

The MapReduce framework contains two main phases, map and reduce, that are controlled by the master machine (i.e., by its driver program). In the map phase, the driver loads the input data, divides it into sub-tasks for the computing nodes (slave nodes named mappers in this phase). The MapReduce driver instructs these nodes to perform calculations according to the user-defined program for mappers (UDPm). The results are saved to intermediate files.

In the reduce phase (after the calculations on mappers have been completed), the driver deploys computing nodes (reducers in this phase) to tally results from the intermediate files. These intermediate results are tallied by a user-defined program for reducers (UDPr) thereby producing report results.

7.2.2 Data Processing Using Hadoop

Apache Hadoop *(http://hadoop.apache.org/)* is a widely used open source implementation of the Google MapReduce [96] distributed data processing framework. It uses the Hadoop distributed file system (HDFS) for data storage across multiple local disks of available computing nodes while presenting a single file system view via an application program interface (API). HDFS is an open source implementation of the Google File System (GFS) [136]. Like GFS, HDFS is typically run on commodity storage equipment. High data availability levels are achieved through data replication. Throughput performance is optimized by scheduling data transfer from a data replica located nearest the computation node. Hadoop performs duplicate executions of slower tasks and handles failures by rerunning failed tasks using different worker nodes.

Hadoop, written in Java, leverages HDFS to reliably store data across multiple computers via a cluster consisting of a single master node and a varying number of slave nodes. The slave nodes can act as the computing nodes for MapReduce and as data nodes for the HDFS. Hadoop is widely used in bioinformatics research. For example, it has been employed to develop algorithms for the analysis of next-generation sequencing data to implement systems for sequence alignment, and develop proteomic search engines [98, 224, 226, 232]. The Hadoop open source implementation is sponsored by Amazon, Yahoo, Google, IBM and other major vendors. Like Google's proprietary MapReduce framework, application developers need only write custom map and reduce functions, and the Hadoop framework automatically executes those functions in parallel. Hadoop and HDFS are used to manage production clusters with 10,000+ nodes and petabytes of data.

Several organizations offer cloud compute cycles that can be accessed via Hadoop. Amazon's Elastic Compute Cloud (EC2) (*http://aws.amazon.com*) contains tens of thousands of virtual machines, and supports Hadoop with minimal effort. Amazon offers preconfigured disk images and launches scripts for initializing Hadoop clusters. Once initialized, users copy data into the newly created HDFS and execute their jobs as if the cluster was dedicated for their use.

As an illustration of solving a problem using a MapReduce application, Figure 7.1 represents a hypothetical application to solve a word counting problem. The objective is to evaluate multiple inputs of arbitrary text and identify the number of times that a certain word occurs in all of the inputs. The result is simply the total number of times that a target word occurs in all inputs. This problem lends itself very well to parallel processing, and by extension, to MapReduce. Each input can be evaluated in parallel, independently and in any order relative to one another, without impacting the final result. This type of problem is said to be commutative, i.e., $A + B = B + A = C$.

The input to a MapReduce application may consist of a large file divided into a number of small chunks or partitions, or it may consist of a large number of independent files. The MapReduce application developer may automate the input preparation as part of its driver function or prepare the input prior to the running of the MapReduce application. For the word counting application illustrated in Figure 7.1 , the inputs are five hypothetical articles from the PubMed database (*http://www.ncbi.nlm.nih.gov/pubmed/*), and the target word is "gene." Each of the five hypothetical PubMed articles cited in Figure 7.1 contains a certain number of occurrences of the word "gene." Each article is represented by one input symbol. Within each of the five input symbols are zero or more text strings: "... gene ..." represents the occurrence of "gene" in that hypothetical article. Hypothetical PubMed articles 1 through 5 are shown to have two, one, one, zero and three occurrences of the word. There are seven occurrences of "gene" in the five hypothetical PubMed articles in total.

The MapReduce framework allows the application developer to decide the number of machines designated to perform the map and reduce functions. In the example in Figure 7.1, there are three machines designated to perform the map function and one machine designated to perform the reduce function. The three machines designated to perform the map function are identified as mapper machines 1 through 3. The power of MapReduce is leveraged best when the number of inputs (files, chunks, partitions, etc.) is larger than the number of mapper machines. The framework is responsible for scheduling available mapper machines as needed to maximize the performance of the MapReduce application.

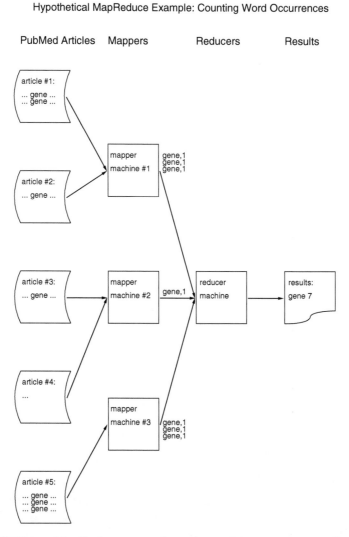

FIGURE 7.1: MapReduce example 1: A simplified "word count" application deployed using the MapReduce framework. The objective is to count and report the number of times the word "gene" appears in a number of PubMed articles. Three mapper processes and one reducer process are shown.

The machine designated to perform the reduce function in Figure 7.1 is identified as reducer machine. In MapReduce, each reducer machine produces a result. This figure shows a single result.

In the course of running a MapReduce application, the framework schedules inputs to be processed by mapper machines. However, the MapReduce application developer need only be concerned about the logic of the map and the reduce functions. In MapReduce, the map function must emit two data elements to be processed by the reduce function. These two data elements are called "key" and "value" and collectively are called a "key-value pair".

In the example in Figure 7.1 , the purpose of the map function is to emit a key-value pair "gene, 1" each time "gene" is found in the input. The purpose of the reduce function in Figure 7.1 is to gather the emissions produced by the mapper machines and tally the number of occurrences. In other words, the reducer simply adds the ones in each key-value pair emitted by the mapper machines.

Figure 7.1 illustrates the passing of key-value pairs from each mapper to the reducer and the result produced by the reducer machine. Since there are more inputs than mapper machines, the MapReduce framework assigns a new input to a mapper machine as soon as the mapper completes the processing of an input. Figure 7.1 illustrates that mapper machines 1 and 2 process two inputs, mapper 3 processes one input and the reducer processes seven key-value pairs.

The MapReduce framework manages most of the low-level details, including data distribution, communication, fault tolerance, etc. The researcher need not be concerned about these details and can therefore concentrate on the algorithms and define the map and reduce routines.

Advances in computational technologies are inundating the biological research community with enormous quantities of valuable and detailed data, while simultaneously supplying the researchers with numerous powerful and useful tools to convert that data into new problem solving intelligence.

A novel tool named Stormbow analyzes RNA-seq samples in the cloud [417]. Amazon Web Services (*http://aws.amazon.com/*) was utilized as the infrastructure for Stormbow. The tool took 6 to 8 hours to process an RNA-seq sample with 100 million reads. The average cost was $3.50 per sample.

The bioinformatics community is challenged by increasingly large data sets required for computational jobs that take unacceptably long times if performed on a small number of machines. The amount of available biological sequence data is growing at a rate described by Moore's law as doubling each 18 months. In other respects, however, the rate of this growth shatters Moore's law [1]. For these cases, distributed comput-

TABLE 7.3: Cloud-based platforms and their URLs

Platform	URL
Amazon Web Services (AWS)	http://aws.amazon.com/
Eoulsan [188]	http://transcriptome.ens.fr/eoulsan/
NIH Biowulf	http://biowulf.nih.gov/
Galaxy Project [139]	http://galaxyproject.org/

ing on multiple clusters at different locations is an attractive approach to achieve short execution times.

Table 7.3 lists cloud-based platforms available for biological data processing. Galaxy [9, 139] is an open, Web-based platform for data-intensive biomedical research. Galaxy is intended to fill the need for re-producible research systems (RRS) whereby a bioinformatics result can be reproduced. To achieve this objective, Galaxy provides management utilities for input data, intermediate results, configuration parameters, etc. for each individual experiment. Galaxy allows the user to perform, reproduce, and share complete analyses on the freely available server or on a private server. CloudMan offers the Galaxy product as a Web service and with Hadoop cluster provisioning capabilities. The Bio-Linux suite of over 250 bioinformatics utilities is available on CloudMan [119]. CloudMan enables a researcher to easily deploy, customize, and share total cloud analysis environment, including data, tools, and configurations with the objective of fulfilling the RRS goals. The CloudMan platform is available for instantiation on the Amazon Web Services (AWS) cloud infrastructure as part of the Cloud Galaxy application and with Cloud-BioLinux (*http://cloudbiolinux.org*) [10].

The CloudBLAST tool runs on a Hadoop platform to perform faster BLAST searching [253]. An Amdahl speed-up of 57 was reported using CloudBLAST on a local Hadoop cluster as compared against a standard BLAST search [13].

Eoulsan is a versatile framework based on the Hadoop implementation of the MapReduce algorithm [188]. Eoulsan is dedicated to high throughput sequencing data analysis on distributed computers. Working either on stand-alone workstations or cloud computing clusters, Eoulsan provides an integrated and flexible solution for analysis of RNA-seq gene expression data, i.e., read mapping. Eoulsan allows users to easily set up a cloud computing cluster on the Amazon Web Services (AWS) public cloud.

Biowulf is a high performance computation (HPC) resource at the National Institutes of Health (NIH). HPC systems at NIH provide computational resources and support for the NIH bioscience research community. Biowulf is a 20,000+ core Linux cluster. A wide variety of scientific software is installed and maintained on Biowulf, along with scientific databases (*http://hpc.nih.gov*).

7.2.3 Next-generation Sequence Data Analysis

Next-generation sequencing (NGS) is defined by a new set of technologies capable of producing sequences of biological products (DNA, RNA and protein) faster and less expensively than previous sequencing technologies [234]. NGS technologies have revolutionized genome research. For instance, a new protocol for sequencing the messenger RNA in a cell known as RNA-seq generates millions of short sequence fragments in a single run. NGS technologies' application to transcriptomics (RNA-seq) is increasingly used for gene expression profiling as a replacement for microarrays. Since the emergence of NGS sequence-based technologies, it has become cost-effective to conduct gene expression profiling research on organisms beyond those considered the model organisms [111].

In microarray analysis, expression levels of large numbers of genes are measured, detected and quantified by detection of fluorophore-labeled or chemiluminescence-labeled targets. Microarray analysis is used to determine relative abundance of nucleic acid sequences in the target in that brightness levels are digitized for computational analysis. In contrast, with NGS analysis, the intention is to provide a physical count of each specific biological product in a sample. Gene expression studies using RNA-seq are sequence-based and offer the potential for much greater analytical capability compared with microarray studies. Having nucleotide visibility allows researchers to detect SNPs, exon splice locations, gene fusion, and other factors that are not possible with microarray analysis [282]. Using NGS, numerous short (50 to 300 nt) "reads" are produced, with each read representing a sequence of nucleic acids. In a gene expression study, the sample analyzed represents the cellular transcriptome from a living organism, and each read represents an expressed gene product, i.e., an RNA transcript. To determine the gene from which the transcript originated, high performance computational (HPC) analysis is used. This determination process is called read mapping. A reference transcription or reference genome is used as a database in a biological search process with the short read being the target for which a match in the database is sought [416].

The properties of RNA-seq data have not been yet fully established, however, and additional research is needed to determine how these data respond to differential expression analysis. Before the potential of RNA-seq can be fully realized, a number of issues need to be addressed, including genome coverage, standardization, sequencing depth, differentiation determination and data normalization [100, 359, 386].

A few issues to be considered in the read mapping process include:

- *Gene length*

  A longer gene will have a larger read-to-gene ratio than a shorter gene. Normalization steps are needed for certain situations such as this. A longer gene is over-represented in a sample since there are more short reads from a longer gene than from a smaller gene. One common approach to this issue is the "reads per kilobase of exon model per million mapped reads" (RPKM) normalization method [282].

- *Exon splice points*

  A gene is comprised of alternating sections called introns and exons. Exons are expressed as RNA and introns are not expressed. A gene transcript may be comprised of multiple exons spliced together. Expressed exons are spliced together in a cell to form a complete messenger RNA (mRNA). Unfortunately, when mapping an RNA-seq read to a reference genome, there will be no match at the splice point. Splicing detection algorithms address the problem of RNA-seq reads failing to match the reference genome [251, 282]. In this case, a reference transcriptome is preferred. Since the field of genomics is still in its infancy few reliable reference transcriptomes are available, even among well studied model organisms.

- *Sample quality*

  A quality score is associated with each RNA-seq sample [282]. This score identifies the resolution and reliability of the sequencing process. Since there is invariably a potential for a missed or mislabeled nucleotide, quality can be improved, for instance, by performing the sequence process multiple times on a read. Although an increased experimental cost is associated with a higher quality score, the higher cost may be justified for certain experiments. On the other hand, there are times when a lower quality score is satisfactory. In general, a higher throughput NGS experiment will produce a different RNA-seq data set compared with a higher resolution NGS experiment [310].

- *Differentiation determination*

  A basic task in the analysis of count data from RNA-seq is the detection of differentially expressed genes. Tools are available to report the number of reads that have been assigned to a gene for each sample [16]. Gene expression is deemed to be differential when there is a measurable gene expression change between two distinct conditions. Otherwise, the expression is deemed non-differential. Experimental standards will help clarify which differential analysis tools are the best performers. Sequencing depth is a factor in determining whether gene expression is differential [310, 339].

- *Sequencing depth and genome coverage*
  The ability to detect and quantify weakly expressed transcripts is challenged by the high concentration of a few highly expressed genes [359]. Increased genome coverage is related to increased read sequencing, which also increases experiment costs, as noted for sample quality described above. Research suggests that sequencing 10 million (M) reads will produce good genome coverage [386].

Cloud-based resources provide massively parallel computing hardware resources at reasonable cost. Simultaneously, several cloud-based bioinformatics applications and resources have been developed recently to help address the challenges of NGS data analysis. CloudBurst [323] is a parallel read mapping algorithm for mapping next-generation sequence RNA-seq data to a reference genome. It reports all alignments for each read up to a user-specified number of differences including both mismatches and indels (insertions and deletions). CloudBurst also filters alignments to report the single best non-ambiguous alignment for each read. CloudBurst runs on the Hadoop open source implementation of the MapReduce distributed programming framework (*http://hadoop.apache.org*). Other cloud-based bioinformatics tools include Crossbow (for identifying single nucleotide polymorphisms (SNPs) on a Hadoop cluster on the AWS public cloud service) [147], Rainbow (for genome-scale sequence data analysis) [418], Stormbow (for read mapping) [417], Myrna (for RNA differential expression analysis) [219], CloVR (for sequence analysis) [20], and CloudAligner (for read mapping) [273].

7.2.4   Gene Expression Analysis

Gene expression analysis using next-generation sequence (NGS) technologies is performed in generally three steps. The first process step is the sequencing of RNA transcripts by high-throughput sequencing technology [69]. The translation of the raw signals coming from the high-throughput sequencing equipment into nucleotide bases for every sequenced read produces RNA-seq. The most common RNA-seq output format generated from this step is called FASTQ and contains short reads along with quality scores assigned to each base. RNA-seq is produced at single-base resolution and can be produced at varying depths of genome coverage, depending on the differential expression analysis need. Experiments have shown that beyond a certain depth coverage, the information gained decreases.

The second process step is the reconstruction of the original genomic sequence. This step is also known as assembly, alignment or mapping. When no reference is available, reads are assembled *de novo*. When a reference genome or transcriptome is available, the reads are aligned to the reference genome or transcriptome. In the alignment process, differences from the reference genome or transcriptome, called genetic variations, are detected and noted for further analysis. In the alignment process, gene expression levels are quantified.

The third process step is the analysis of the results. The analysis performed depends on the research objective. In Chapter 6, for example, several methods to infer a gene regulatory network (GRN) are presented and discussed. A GRN is inferred through a reverse engineering procedure by evaluating gene expression profile data and determining the mechanics of the cellular system that produced the gene expression profile. GRN results to date, while not great, are promising. Much has been accomplished in this field but the field is still very young. This potential ability to have a front row seat to view each minute detail of cellular activity is unprecedented.

With next generation sequencing (NGS), much information can be gathered about quantities of gene expressions within cells. Read mapping technologies help quantify gene expression levels. Tools to perform sequence alignment in the read mapping of gene expression analysis include: Bowtie [218], Bowtie2 [220], Soap3 [241], Omicsoft Sequence Aligner (OSA) [172], and TopHat [368]. Sampling gene expressions at periodic time intervals helps shed light on patterns. Each bit of knowledge about patterns contributes to the understanding of the logic behind this expression activity. A cell may have multiple causes for a certain gene to express RNA (i.e., mRNA or ncRNA). Such causes may be exclusive and independent from one another, they may all be prerequisites for gene expression, or there may be combinations of causes. Knowledge of this expression logic is valuable for researchers. For statistical tests, several R-based packages are used specifically for differential expression (DE) analysis in RNA-seq experiments, including DESeq [16] and edgeR [17]. In a comparison of several methods to analyze RNA-seq differential expression, DESeq and edgeR ranked among the best performers [339].

As an illustration of solving a biological problem using a MapReduce application, Figure 7.2 represents a hypothetical application to evaluate and quantify gene expressions. The objectives were to evaluate multiple inputs of RNA-seq transcripts, i.e., reads, expressed from hypothetical genes, determine the hypothetical gene that expressed each read, and identify the number of times that each hypothetical gene was expressed in the RNA-seq transcriptome dataset. The results reveal the total number of times that all hypothetical genes were expressed based on all inputs. Gene expression quantification is a very active research topic in bioinformatics. Gene expression levels can be deduced from the total number of reads that fall into the transcript expression regions or exons of a gene [387] . As NGS technologies continue to improve, RNA-seq genomic data will become less expensive, and the production of these data will outpace the abilities of the medical community to analyze them. It is important to develop new bioinformatics analysis tools to ensure the timely analysis of this wealth of data.

This hypothetical gene expression counting problem, like the previous word counting MapReduce example illustrated in Figure 7.1 , lends itself very well to parallel processing, and by extension, to MapReduce. Each RNA-seq input can be evaluated in parallel, independently and in an order relative to one another, without impacting the final result.

For the gene expression counting MapReduce application illustrated in Figure 7.2 , the inputs are five partitions of a hypothetical RNA-seq transcriptome dataset. For this hypothetical application, the objective was to quantify the expression levels of four hypothetical genes, called G1, G2, G3 and G4. Table 7.4 illustrates hypothetical nucleotide sequences associated with these genes. Each of the five partitions of the hypothetical RNA-seq transcriptome dataset in Figure 7.2 contains a certain number of occurrences of nucleotide sequences from Table 7.4. Each of the five partitions is represented by one input symbol. Within each of the five input symbols are one or more text strings, e.g. "... acg ...", which represent the occurrences of a nucleotide sequence in that portion of the dataset. In Figure 7.2 , hypothetical RNA-seq partitions 1 through 5 are shown to have one, one, two, one and three, respectively, instances of a nucleotide sequence expressed from the four hypothetical genes of interest. Note that there is no gene identification information represented in any RNA-seq transcriptome dataset. The first task of any RNA-seq data analysis is to compare the nucleotide sequences, i.e., reads, from the RNA-seq transcriptome dataset to the nucleotides in a complete and annotated genome, i.e., reference genome. The reference genome is the complete and fully annotated genome of the organism from which the RNA-seq transcriptome dataset was derived. The process of comparing a read to a reference genome is called mapping. By mapping a nucleotide sequence to a reference genome, the gene that expressed the nucleotide sequence can be identified. In the five inputs shown in Figure 7.2 eight instances of a nucleotide sequence were expressed from the four hypothetical genes of interest, as shown in Table 7.4 .

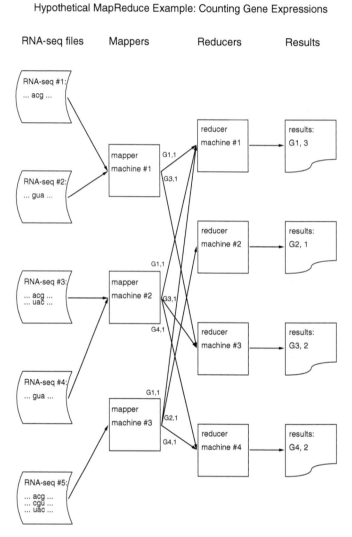

FIGURE 7.2: MapReduce example 2: A simplified gene expression counting application deployed using the MapReduce framework. The objective is to count and report the number of times each gene is expressed in a collection of RNA-seq data files. Three mapper processes and four reducer processes are shown.

TABLE 7.4: Hypothetical gene-ID RNA-seq list

Gene	RNA-seq Read Sequence
G1	acg
G2	cgu
G3	gua
G4	uac

NOTE: An RNA nucleotide sequence is shown for each of four hypothetical genes used in a MapReduce example in the chapter.

The MapReduce framework allows the application developer to decide the number of machines designated to perform the map and reduce functions. These values can be selected to balance performance and budgetary constraints. In the MapReduce example illustrated in Figure 7.2, three machines are designated to perform the map function and four machines designated to perform the reduce function. The three machines designated to perform the map function are identified as mapper machines 1 through 3. The power of MapReduce is leveraged best when the number of inputs (files, chunks, partitions, etc.) is larger than the number of mapper machines. The MapReduce framework is responsible for scheduling available mapper machines as needed to maximize the performance.

The four machines designated to perform the reduce function in Figure 7.2 are identified as reducer machines 1 through 4. In MapReduce, each reducer machine produces a result. Therefore, this figure shows four results.

As described in Section 7.2.2, in the course of a running MapReduce application, the framework schedules inputs to be processed by mapper machines. However, the application developer need only be concerned about the logic of the map and reduce functions. Since the map function must emit a key-value pair to be processed by the reduce function for the example in Figure 7.2, this key-value pair consists of Gn, 1, where n is 1, 2, 3 or 4, and Gn represents a hypothetical gene from the list in Table 7.4 each time a corresponding RNA-seq read sequence is found in the input. The purpose of the reduce function in Figure 7.2 is to gather the emissions produced by the mapper machines and tally the number of occurrences. In other words, the reducer machine simply adds up the ones in each key-value pair emitted by the mapper machines.

Figure 7.2 illustrates the passing of key-value pairs from each mapper to the reducer and the result produced by the reducer machine. Since there are more inputs than mapper machines, the MapReduce framework assigns a new input to a mapper as soon as that mapper completes the processing of an input. Figure 7.2 illustrates that mapper machines 1 and 2 each process two inputs, mapper 3 processes one input, reducer 1 processes 3 key-value pairs, reducer 2 processes 1 key-value pair and reducers 3 and 4 each process two key-value pairs.

Bibliography

[1] Human genome at ten: The sequence explosion. *Nature*, 464(7289):670–671, 2010.

[2] D. P. Aalberts and N. O. Hodas. Asymmetry in RNA pseudoknots: observation and theory. *Nucleic Acids Research*, 33:2210–2214, 2005.

[3] D. P. Aalberts and N. Nandagopal. A two-length-scale polymer theory for RNA loop free energies and helix stacking. *RNA*, 16:1350–1355, 2010.

[4] D. Abdulrehman, P. T. Monteiro, M. C. Teixeira, N. P. Mira, A. B. Lourenço, S. C. dos Santos, T. R. Cabrito, A. P. Francisco, S. C. Madeira, R. S. Aires, A. L. Oliveira, I. Sá-Correia, and A. T. Freitas. YEASTRACT: providing a programmatic access to curated transcriptional regulatory associations in *Saccharomyces cerevisiae* through a Web services interface. *Nucleic Acids Research*, 39:D136–D140, 2011.

[5] E. Acerbi, T. Zelante, V. Narang, and F. Stella. Gene network inference using continuous time Bayesian networks: a comparative study and application to Th17 cell differentiation. *BMC Bioinformatics*, 15, 2014.

[6] F. Achard, G. Vaysseix, and E. Barillot. XML, bioinformatics and data integration. *Bioinformatics*, 17:115–125, 2001.

[7] A. Acland, R. Agarwala, T. Barrett, J. Beck, D. A. Benson, C. Bollin, E. Bolton, S. H. Bryant, K. Canese, D. M. Church, K. Clark, M. DiCuccio, I. Dondoshansky, S. Federhen, M. Feolo, L. Y. Geer, V. Gorelenkov, M. Hoeppner, M. Johnson, C. Kelly, V. Khotomlianski, A. Kimchi, M. Kimelman, P. Kitts, S. Krasnov, A. Kuznetsov, D. Landsman, D. J. Lipman, Z. Lu, T. L. Madden, T. Madej, D. R. Maglott, A. Marchler-Bauer, I. Karsch-Mizrachi, T. Murphy, J. Ostell, C. O'Sullivan, A. Panchenko, L. Phan, D. Preussm, K. D. Pruitt, W. Rubinstein, E. W. Sayers, V. Schneider, G. D. Schuler, E. Sequeira, S. T. Sherry, M. Shumway,

K. Sirotkin, K. Siyan, D. Slotta, A. Soboleva, G. Starchenko, T. A. Tatusova, B. Trawick, D. Vakatov, Y. Wang, M. W. Ward, W. J. Wilbur, E. Yaschenko, and Zbic. Database resources of the National Center for Biotechnology Information. *Nucleic Acids Research*, 41(D1):D8–D20, 2013.

[8] P. L. Adams, M. R. Stahley, A. B. Kosek, J. Wang, and S. A. Strobel. Crystal structure of a self-splicing group I intron with both exons. *Nature*, 430(6995):45–50, 2004.

[9] E Afgan, D Baker, N Coraor, B Chapman, A Nekrutenko, and J Taylor. Galaxy CloudMan: delivering cloud compute clusters. *BMC Bioinformatics*, 11, 2010.

[10] E. Afgan, B. Chapman, M. Jadan, V. Franke, and J. Taylor. Using cloud computing infrastructure with CloudBioLinux, CloudMan, and Galaxy. *Current Protocols in Bioinformatics*, (38), 2012.

[11] K. Aggarwal and K. H. Lee. Overexpression of cloned RhsA sequences perturbs the cellular translational machinery in *escherichia coli*. *Journal of Bacteriology*, 193(18):4869–4880, 2011.

[12] F. H. Allen, S. Bellard, M. D. Brice, B. A. Cartwright, A. Doubleday, H. Higgs, T. Hummelink, B. G. Hummelink-Peters, O. Kennard, W. D. S. Motherwell, J. R. Rodgers, and D. G. Watson. The Cambridge Crystallographic Data Centre: computer-based search, retrieval, analysis and display of information. *Acta Crystallographica Section B*, 35(10):2331–2339, 1979.

[13] S. F. Altschul, W. Gish, W. Miller, E. W. Myers, and D. J. Lipman. Basic local alignment search tool. *Journal of Molecular Biology*, 215:403–410, 1990.

[14] S. F. Altschul, T. L. Madden, A. A. Schaffer, J. Zhang, Z. Zhang, W. Miller, and D. J. Lipman. Gapped BLAST and PSI-BLAST: a new generation of protein database search programs. *Nucleic Acids Research*, 25:3389–3402, 1997.

[15] G. M. Amdahl. Validity of the single processor approach to achieving large scale computing capabilities. In *Proceedings of April 18-20, 1967, Spring Joint Computer Conference*, pages 483–485, New York, 1967. ACM.

[16] S. Anders and W. Huber. Differential expression analysis for sequence count data. *Genome Biology*, 11(10), 2010.

[17] S. Anders, W. Huber, D. J. McCarthy, Y. Chen, G. K. Smyth, M. Okoniewski, and M. D. Robinson. Count-based differential expression analysis of RNA sequencing data using R and Bioconductor. *Nature Protocols*, 8(9):1765–1786, 2013.

[18] M. Andronescu, V. Bereg, H. H. Hoos, and A. Condon. RNA STRAND: the rna secondary structure and statistical analysis database. *BMC Bioinformatics*, 9, 2008.

[19] I. P. Androulakis, E. Yang, and R. R. Almon. Analysis of time-series gene expression data: methods, challenges, and opportunities. *Annual Review of Biomedical Engineering*, 9:205–228, 2007.

[20] S. V. Angiuoli, M. Matalka, A. Gussman, K. Galens, M. Vangala, D. R. Riley, C. Arze, J. R. White, O. White, and W. F. Fricke. CloVR: A virtual machine for automated and portable sequence analysis from the desktop using cloud computing. *BMC Bioinformatics*, 12, 2011.

[21] S. V. Angiuoli, J. R. White, M. Matalka, O. White, and W. Fricke. Resources and costs for microbial sequence analysis evaluated using virtual machines and cloud computing. *PLoS One*, 6(10), 2011.

[22] S. E. Antonarakis and V. A. McKusick. OMIM passes the 1,000-disease-gene mark. *Nature Genetics*, 25(1):11–11, 2000.

[23] M. Armbrust, A. Fox, R. Griffith, A. D. Joseph, R. H. Katz, A. Konwinski, G. Lee, D. A. Patterson, A. Rabkin, I. Stoica, and M. Zaharia. Above the Clouds: A Berkeley View of Cloud Computing. Technical report, University of California, Berkeley, 2009.

[24] M. Armbrust, R. S. Xin, C. Lian, Y. Huai, D. Liu, J. K. Bradley, X. Meng, M. J. Frankliny, A. Ghodsi, M. Zaharia, and T. Kaftan. Spark SQL: Relational data processing in spark. In *Proceedings of ACM SIGMOD International Conference on Management of Data*, volume 2015-May, pages 1383–1394, Databricks Inc., 2015.

[25] Y. Asakura, H. Kojima, and I. Kobayashi. Evolutionary genome engineering using a restriction-modification system. *Nucleic Acids Research*, 39(20):9034–9046, 2011.

[26] M. Ashburner, C. A. Ball, J. A. Blake, D. Botstein, H. Butler, J. M. Cherry, A. P. Davis, K. Dolinski, S. S. Dwight, J. T. Eppig, M. A. Harris, D. P. Hill, L. Issel-Tarver, A. Kasarskis, S. Lewis, J. C. Matese, J. E. Richardson, M. Ringwald, G. M. Rubin, and G. Sherlock. Gene Ontology: tool for the unification of biology. *Nature Genetics*, 25(1):25–29, 2000.

[27] S. Baba, K. I. Takahashi, S. Noguchi, H. Takaku, G. Kawai, Y. Koyanagi, and N. Yamamoto. Solution RNA structures of the HIV-1 dimerization initiation site in the kissing-loop and extended-duplex dimers. *Journal of Biochemistry*, 138(5):583–592, 2005.

[28] M. S. Babcock and W. K. Olson. A new program for the analysis of nucleic acid structure: Implications for nucleic acid structure interpretation. In D. M. Soumpasis and T. M. Jovin, editors, *Computation of Biomolecular Structures*, pages 65–85. Springer, 1993.

[29] A. Bairoch. The ENZYME database in 2000. *Nucleic Acids Research*, 28:304–305, 2000.

[30] N. Ban, P. Nissen, J. Hansen, P. B. Moore, and T. A. Steitz. The complete atomic structure of the large ribosomal subunit at 2.4 A resolution. *Science*, 289:905–920, 2000.

[31] Z. Bar-Joseph. Analyzing time series gene expression data. *Bioinformatics (Oxford, England)*, 20(16):2493–2503, 2004.

[32] T. Barrett and R. Edgar. Mining microarray data at NCBI's gene expression omnibus (GEO)*. *Methods in Molecular Biology*, 338:175–190, 2006.

[33] T Barrett, SE Wilhite, P Ledoux, C Evangelista, IF Kim, M Tomashevsky, KA Marshall, KH Phillippy, PM Sherman, M Holko, A Yefanov, H Lee, N Zhang, CL Robertson, N Serova, S Davis, and A Soboleva. NCBI GEO: archive for functional genomics data sets–update. *Nucleic Acids Research*, 41:D991–D995, 2013.

[34] A. Bateman, E. Birney, L. Cerruti, R. Durbin, L. Etwiller, S. R. Eddy, S. Griffiths-Jones, K. L. Howe, M. Marshall, and E. L. L. Sonnhammer. The Pfam protein families database. *Nucleic Acids Research*, 30:276–280, 2002.

[35] A. T. Belew, A. Meskauskas, S. Musalgaonkar, V. M. Advani, S. O. Sulima, J. D. Dinman, W. K. Kasprzak, and B. A. Shapiro. Ribosomal frameshifting in the CCR5 mRNA is regulated by miRNAs and the NMD pathway. *Nature*, 512(7514):265–269, 2014.

[36] P. Bellot, P. Salembier, A. Oliveras-Vergés, P. E. Meyer, and C. Olsen. NetBenchmark: a bioconductor package for reproducible benchmarks of gene regulatory network inference. *BMC Bioinformatics*, 16(1), 2015.

[37] D. A. Benson, M. Cavanaugh, K. Clark, I. Karsch-Mizrachi, D. J. Lipman, J. Ostell, and E. W. Sayers. GenBank. *Nucleic Acids Research*, 41:D36–D42, 2013.

[38] H. M. Berman, J. Westbrook, Z. Feng, G. Gilliland, T. N. Bhat, H. Weissig, I. N. Shindyalov, and P. E. Bourne. The protein data bank. *Nucleic Acids Research*, 28:235–242, 2000.

[39] V. Berry and D. Bryant. Faster reliable phylogenetic analysis. In *Proceedings of Annual International Conference on Computational Molecular Biology*, pages 59–68, Universite de Saint-Etienne, 1999.

[40] T. N. Bhat, G. Gilliland, V. Ravichandran, N. Thanki, P. Bourne, H. Weissig, Z. Feng, S. Jain, B. Schneider, K. Schneider, J. Westbrook, and H. M. Berman. The PDB data uniformity project. *Nucleic Acids Research*, 29(1):214–218, 2001.

[41] D. Bitton and D. J. DeWitt. Duplicate record elimination in large data files. *ACM Transactions on Databases*, 8:255–265, 1983.

[42] J. A. Blake, M. Dolan, H. Drabkin, D. P. Hill, L. Ni, D. Sitnikov, S. Burgess, T. Buza, C. Gresham, F. McCarthy, L. Pillai, H. Wang, S. Carbon, S. E. Lewis, C. J. Mungall, P. Gaudet, R. L. Chisholm, P. Fey, W. A. Kibbe, S. Basu, D. A. Siegele, B. K. McIntosh, D. P. Renfro, A. E. Zweifel, J. C. Hu, N. H. Brown, S. Tweedie, Y. Alam-Faruque, R. Apweiler, A. Auchincloss, K. Axelsen, G. Argoud-Puy, B. Bely, M. C. Blatter, L. Bougueleret, E. Boutet, S. Branconi-Quintaje, L. Breuza, A. Bridge, P. Browne, W. M. Chan, E. Coudert, I. Cusin, E. Dimmer, P. Duek-Roggli, R. Eberhardt, A. Estreicher, L. Famiglietti, S. Ferro-Rojas, M. Feuermann, M. Gardner, A. Gos, N. Gruaz-Gumowski, U. Hinz, C. Hulo, R. Huntley, J. James, S. Jimenez, F. Jungo, G. Keller, K. Laiho, D. Legge, P. Lemercier, D. Lieberherr, M. Magrane, M. J. Martin, P. Masson, M. Moinat, C. O'Donovan, I. Pedruzzi, K. Pichler, D. Poggioli, P. Porras Millán, S. Poux, C. Rivoire, B. Roechert, T. Sawford, M. Schneider, H. Sehra, E. Stanley, A. Stutz, S. Sundaram, M. Tognolli, I. Xenarios, R. Foulger, J. Lomax, P. Roncaglia, E. Camon, V. K. Khodiyar, R. C. Lovering, P. J. Talmud, M. Chibucos, M. Gwinn Giglio, K. Dolinski, S. Heinicke, M. S. Livstone, R. Stephan, M. A. Harris, S. G. Oliver, K. Rutherford, V. Wood, J. Bahler, A. Lock, P. J. Kersey, M. D. McDowall, D. M. Staines, M. Dwinell, M. Shimoyama, S. Laulederkind, T. Hayman, S. J. Wang, V. Petri, T. Lowry, P. D'Eustachio, L. Matthews, C. D. Amundsen, R. Balakrishnan, G. Binkley, J. M. Cherry, K. R. Christie, M. C. Costanzo, S. S. Dwight, S. R. Engel, D. G. Fisk, J. E. Hirschman, B. C. Hitz, E. L. Hong, K. Karra, C. J. Krieger, S. R. Miyasato, R. S. Nash, J. Park, M. S. Skrzypek, S. Weng, E. D. Wong, T. Z. Berardini, D. Li, E. Huala, D. Slonim, H. Wick, P. Thomas, J. Chan, R. Kishore, P. Sternberg,

K. Van Auken, D. Howe, and M. Westerfield. Gene Ontology: enhancements for 2011. *Nucleic Acids Research*, 40(D1):D559–D564, 2012.

[43] J. A. Blake and M. A. Harris. Gene Ontology (GO) project: structured vocabularies for molecular biology and their application to genome and expression analysis. *Current Protocols in Bioinformatics*, Chapter 7, 2008.

[44] F. R. Blattner, G. Plunkett, C. A. Bloch, N. T. Perna, V. Burland, M. Riley, J. Collado-Vides, J. D. Glasner, C. K. Rode, G. F. Mayhew, J. Gregor, N. W. Davis, H. A. Kirkpatrick, M. A. Goeden, D. J. Rose, B. Mau, and Y. Shao. The complete genome sequence of *Escherichia coli* K-12. *Science*, (5331), 1997.

[45] M. S. Boguski, T. M. Lowe, and C. M. Tolstoshev. dbEST: database for expressed sequence tags. *Nature Genetics*, 4:332–333, 1993.

[46] H. Bolouri. *Computational Modeling of Gene Regulatory Networks : A Primer.* Imperial College Press, 2008.

[47] Richard Bonneau, David J Reiss, Paul Shannon, Marc Facciotti, Leroy Hood, Nitin S Baliga, and Vesteinn Thorsson. The Inferelator: an algorithm for learning parsimonious regulatory networks from systems biology data sets de novo. *Genome Biology*, 7(5), 2006.

[48] C. M. Bouton and J. Pevsner. DRAGON: database referencing of array genes online. *Bioinformatics*, 16(11):1038–1039, 2000.

[49] H. Boutselakis, D. Dimitropoulos, J. Fillon, A. Golovin, K. Henrick, A. Hussain, J. Ionides, M. John, P. A. Keller, E. Krissinel, P. McNeil, A. Naim, R. Newman, T. Oldfield, J. Pineda, A. Rachedi, J. Copeland, A. Sitnov, S. Sobhany, A. Suarez-Uruena, J. Swaminathan, M. Tagari, J. Tate, S. Tromm, S. Velankar, and W. Vranken. E-MSD: the European Bioinformatics Institute Macromolecular Structure Database. *Nucleic Acids Research*, 31(1):458–462, 2003.

[50] L. Breiman. Random forests. *Machine Learning*, 45:5–32, 2001.

[51] L. Breiman, J. H. Friedman, R. A. Olshen, and C. J. Stone. *Classification and Regression Trees.* Wadsworth Belmont, CA, 1984.

[52] R. Breitling, M. L. Stewart, M. P. Barrett, S. Ritchie, and D. Goodenowe. *Ab initio* prediction of metabolic networks using Fourier transform mass spectrometry data. *Metabolomics*, 2(3):155–164, 2006.

[53] A. Bremges, S. Schirmer, and R. Giegerich. Fine-tuning structural RNA alignments in the twilight zone. *BMC Bioinformatics*, 11, 2010.

[54] G. S. Brodal, R. Fagerberg, and C. N. S. Pedersen. Computing the quartet distance between evolutionary trees in time O(nlog2n). In *Proceedings of 12th Annual International Symposium on Algorithms and Computation.*, pages 731–742, 2001.

[55] E. K. Brown and W. H. E. Day. A computationally efficient approximation to the nearest neighbor interchange metric. *Journal of Classification*, 1:93–124, 1984.

[56] J. W. Brown. The Ribonuclease P database. *Nucleic Acids Research*, 27(1):351–352, 1999.

[57] S. Brunak, A. Danchin, M. Hattori, H. Nakamura, K. Shinozaki, T. Matise, and D. Preuss. Nucleotide sequence database policies. *Science*, (5597), 2002.

[58] W. J. Bruno, N. D. Socci, and A. L. Halpern. Weighted neighbor joining: a likelihood-based approach to distance-based phylogeny reconstruction. *Molecular Biology and Evolution*, 17:189–197, 2000.

[59] D. Bryant, J. Tsang, P. Kearney, and M. Li. Computing the quartet distance between evolutionary trees. In *Proceedings of 11th Annual ACMSIAM Symposium on Discrete Algorithms.*, 2000.

[60] P. Buneman. The recovery of trees from measures of dissimilarity. *Mathematics in Archaeological and Historical Sciences*, pages 387–395, 1971.

[61] S. W. Burge, J. Daub, R. Eberhardt, J. Tate, L. Barquist, A. Bateman, E. P. Nawrocki, S. R. Eddy, and P. P. Gardner. Rfam 11.0: 10 years of RNA families. *Nucleic Acids Research*, 41(D1):D226–D232, 2013.

[62] K. Byron, M. Cervantes-Cervantes, J. T. L. Wang, W. C. Lin, and Y. Park. Mining *roX1* RNA in *Drosophila* genomes using covariance models. *International Journal of Computational Bioscience*, 1(1):22–32, 2010.

[63] A. Calf, G. De Giacomo, and M. Lenzerini. Models for information integration: turning local-as-view into global-as-view. In *Proceedings of International Workshop on Foundations of Models for Information Integration*, 2001.

[64] A. Calì. Reasoning in data integration systems: why LAV and GAV are siblings. In *Proceedings of International Symposium on Methodologies for Intelligent Systems*, pages 562–571, 2003.

[65] J. H. Camin and R. R. Sokal. A method for deducing branching sequences in phylogeny. *Evolution*, 19:311–326, 1965.

[66] E. Capriotti and M. A. Marti-Renom. SARA: a server for function annotation of RNA structures. *Nucleic Acids Research*, 37:W260–W265, 2009.

[67] G. Cardona, F. Rosselló, and G. Valiente. Extended newick: it is time for a standard representation of phylogenetic networks. *BMC Bioinformatics*, 9, 2008.

[68] F. Caruso, M. Cochinwala, A. Ganapathy, G. Lalk, and P. Missier. Telcordia's database reconciliation and data quality analysis tool. In *Proceedings of 26th International Conference on Very Large Data Bases*, pages 615–618, 2000.

[69] F. Casals, Y. Idaghdour, J. Hussin, and P. Awadalla. Next-generation sequencing approaches for genetic mapping of complex diseases. *Journal of Neuroimmunology*, 248(1-2):10–22, 2012.

[70] L. Cerulo, C. Elkan, and M. Ceccarelli. Learning gene regulatory networks from only positive and unlabeled data. *BMC Bioinformatics*, 11, 2010.

[71] V. Chandrasekaran, N. Srebro, and P. Harsha. Complexity of inference in graphical models. In *Proceedings of 24th Conference on Uncertainty in Artificial Intelligence*, pages 70–78, Massachusetts Institute of Technology, 2008.

[72] C-. C. Chang and C-. J. Lin. LIBSVM: A library for support vector machines. *ACM Transactions on Intelligent Systems and Technology*, 2(3):1–27, May 2011.

[73] Y. Chang, J. W. Gray, and C. J. Tomlin. Exact reconstruction of gene regulatory networks using compressive sensing. *BMC Bioinformatics*, 15(1), 2014.

[74] Y. F. Chang, Y. L. Huang, and C. L. Lu. SARSA: a Web tool for structural alignment of RNA using a structural alphabet. *Nucleic Acids Research*, 36:W19–W24, 2008.

[75] N. Chen, T. W. Harris, I. Antoshechkin, C. Bastiani, T. Bieri, D. Blasiar, K. Bradnam, P. Canaran, J. Chan, C. Chen, W. J. Chen, F. Cunningham, P. Davis, E. Kenny, R. Kishore, D. Lawson, R. Lee, H. Muller, C. Nakamura, S. Pai, P. Ozersky, A. Petcherski, A. Rogers, A. Sabo, E. M. Schwarz, K. Van Auken, Q. Wang, R. Durbin, J. Spieth, P. W. Sternberg, and Lincoln D. Stein. WormBase: a comprehensive data resource for *Caenorhabditis* biology and genomics. *Nucleic Acids Research*, 33:D383–D389, 2005.

[76] J. M. Cherry, C. Adler, C. Ball, S. A. Chervitz, S. S. Dwight, E. T. Hester, Y. Jia, G. Juvik, T. Roe, M. Schroeder, S. Weng, and D. Botstein. SGD: *Saccharomyces* genome database. *Nucleic Acids Research*, 26:73–79, 1998.

[77] K. H. Cheung, K. White, J. Hager, M. Gerstein, V. Reinke, K. Nelson, P. Masiar, R. Srivastava, Y. Li, J. Li, H. Zhao, J. Li, D. B. Allison, M. Snyder, P. Miller, and K. Williams. YMD: a microarray database for large-scale gene expression analysis. *Proceedings of American Medical Informatics Association Annual Symposium*, pages 140–144, 2002.

[78] S. L. Chin, I. M. Marcus, R. R. Klevecz, and C. M. Li. Dynamics of oscillatory phenotypes in *Saccharomyces cerevisiae* reveal a network of genome-wide transcriptional oscillators. *The FEBS Journal*, 279(6):1119–1130, 2012.

[79] J. M. Churko, G. L. Mantalas, M. P. Snyder, and J. C. Wu. Overview of high throughput sequencing technologies to elucidate molecular pathways in cardiovascular diseases. *Circulation Research*, 112:1613–1623, 2013.

[80] M. Cochinwala, V. Kurien, G. Lalk, and D. Shasha. Efficient data reconciliation. *Information Sciences*, 137(1-4):1–15, 2001.

[81] G. Cochrane, P. Aldebert, N. Althorpe, M. Andersson, W. Baker, A. Baldwin, K. Bates, S. Bhattacharyya, P. Browne, A. van den Broek, M. Castro, K. Duggan, R. Eberhardt, N. Faruque, J. Gamble, C. Kanz, T. Kulikova, C. Lee, R. Leinonen, Q. Lin, V. Lombard, R. Lopez, M. McHale, H. McWilliam, G. Mukherjee, F. Nardone, M. P. Pastor, S. Sobhany, P. Stoehr, K. Tzouvara, R. Vaughan, D. Wu, W. Zhu, and R. Apweiler. EMBL nucleotide

sequence database: developments in 2005. *Nucleic Acids Research*, 34:D10–D15, 2006.

[82] E. F. Codd. Relational model of data for large shared data banks. *Communications of ACM*, 13(6):377–387, 1970.

[83] R. Cole, M. Farach-Colton, R. Hariharan, T. M. Przytycka, and M. Thorup. An O(nlogn) algorithm for the maximum agreement subtree problem for binary trees. *SIAM Journal on Computing*, 30(5):1385–1404, 2000.

[84] FlyBase Consortium. FlyBase: the *Drosophila* database. *Nucleic Acids Research*, 22(17):3456–3458, 1994.

[85] Gene Ontology Consortium. Gene Ontology: tool for the unification of biology. *Nature Genetics*, 25:25–29, 2000.

[86] Sea Urchin Genome Sequencing Consortium, E. Sodergren, G. M. Weinstock, E. H. Davidson, R. A. Cameron, R. A. Gibbs, R. C. Angerer, L. M. Angerer, M. I. Arnone, D. R. Burgess, R. D. Burke, J. A. Coffman, M. Dean, M. R. Elphick, C. A. Ettensohn, K. R. Foltz, A. Hamdoun, R. O. Hynes, W. H. Klein, W. Marzluff, D. R. McClay, R. L. Morris, A. Mushegian, J. P. Rast, L. C. Smith, M. C. Thorndyke, V. D. Vacquier, G. M. Wessel, G. Wray, L. Zhang, C.G. Elsik, O. Ermolaeva, W. Hlavina, G. Hofmann, P. Kitts, M. J. Landrum, A. J. Mackey, D. Maglott, G. Panopoulou, A. J. Poustka, K. Pruitt, V. Sapojnikov, X. Song, A. Souvorov, V. Solovyev, Z. Wei, C. A. Whittaker, K. Worley, K. J. Durbin, Y. Shen, O. Fedrigo, D. Garfield, R. Haygood, A. Primus, R. Satija, T. Severson, M. L. Gonzalez-Garay, A. R. Jackson, A. Milosavljevic, M. Tong, C. E. Killian, B. T. Livingston, F. H. Wilt, N. Adams, R. Belle, S. Carbonneau, R. Cheung, P. Cormier, B. Cosson, J. Croce, A. Fernandez-Guerra, A. M. Geneviere, M. Goel, H. Kelkar, J. Morales, O. Mulner-Lorillon, A. J. Robertson, J. V. Goldstone, B. Cole, D. Epel, B. Gold, M. E. Hahn, M. Howard-Ashby, M. Scally, J. J. Stegeman, E. L. Allgood, J. Cool, K. M. Judkins, S. S. McCafferty, A. M. Musante, R. A. Obar, A. P. Rawson, B. J. Rossetti, I. R. Gibbons, M. P. Hoffman, A. Leone, S. Istrail, S. C. Materna, M. P. Samanta, V. Stolc, W. Tongprasit, Q. Tu, K. F. Bergeron, B. P. Brandhorst, J. Whittle, K. Berney, D.J. Bottjer, C. Calestani, K. Peterson, E. Chow, Q. A. Yuan, E. Elhaik, D. Graur, J. T. Reese, I. Bosdet, S. Heesun, M. A. Marra, J. Schein, M. K. Anderson, V. Brockton, K. M. Buckley, A. H. Cohen, S. D. Fugmann, T. Hibino, M. Loza-Coll, A. J. Majeske, C. Messier, S. V. Nair,

Z. Pancer, D. P. Terwilliger, C. Agca, E. Arboleda, N. Chen, A. M. Churcher, F. Hallbook, G. W. Humphrey, M. M. Idris, T. Kiyama, S. Liang, D. Mellott, X. Mu, G. Murray, R. P. Olinski, F. Raible, M. Rowe, J. S. Taylor, K. Tessmar-Raible, D. Wang, K. H. Wilson, S. Yaguchi, T. Gaasterland, B. E. Galindo, H. J. Gunaratne, C. Juliano, M. Kinukawa, G. W. Moy, A. T. Neill, M. Nomura, M. Raisch, A. Reade, M. M. Roux, J. L. Song, Y. H. Su, I.K. Townley, E. Voronina, J. L. Wong, G. Amore, M. Branno, E. R. Brown, V. Cavalieri, V. Duboc, L. Duloquin, C. Flytzanis, C. Gache, F. Lapraz, T. Lepage, A. Locascio, P. Martinez, G. Matassi, V. Matranga, R. Range, F. Rizzo, E. Rottinger, W. Beane, C. Bradham, C. Byrum, T. Glenn, S. Hussain, G. Manning, E. Miranda, R. Thomason, K. Walton, A. Wikramanayke, S. Y. Wu, R. Xu, C. T. Brown, L. Chen, R. F. Gray, P. Y. Lee, J. Nam, P. Oliveri, J. Smith, D. Muzny, S. Bell, J. Chacko, A. Cree, S. Curry, C. Davis, H. Dinh, S. Dugan-Rocha, J. Fowler, R. Gill, C. Hamilton, J. Hernandez, S. Hines, J. Hume, L. Jackson, A. Jolivet, C. Kovar, S. Lee, L. Lewis, G. Miner, M. Morgan, L. V. Nazareth, G. Okwuonu, D. Parker, L. L. Pu, R. Thorn, and R. Wright. The genome of the sea urchin *Strongylocentrotus purpuratus*. *Science*, (5801):941–952, 2006.

[87] G. F. Cooper. The computational complexity of probabilistic inference using Bayesian belief networks. *Artificial Intelligence*, 42(2-3):393–405, 1990.

[88] F. F. Costa. Non-coding RNAs: lost in translation? *Gene*, 386(1-2):1–10, 2007.

[89] K. Darty, A. Denise, and Y. Ponty. VARNA: interactive drawing and editing of the RNA secondary structure. *Bioinformatics*, 25(15):1974–1975, 2009.

[90] B. DasGupta, X. He, T. Jiang, M. Li, J. Tromp, L. Wang, and L. Zhang. *Computing Distances between Evolutionary Trees.*, volume 2, pages 35–76. Kluwer Academic, 1998.

[91] T. Dasu and T. Johnson. Problems, solutions and research in data quality. In *Proceedings of SIAM International Conference on Data Mining*, 2002.

[92] T. Dasu and T. Johnson. *Exploratory Data Mining and Data Cleaning*. John Wiley & Sons, 2003.

[93] B. Davidson, C. Overton, and P. Buneman. Challenges in integrating biological data sources. *Journal of Computational Biology*, 2:557–572, 1995.

[94] S. B. Davidson, V. Tannen, C. Overton, and L. Wong. Biokleisli: a digital library for biomedical researchers. *International Journal on Digital Libraries*, 1(1):36–53, 1997.

[95] W. H. E. Day. Optimal algorithms for comparing trees with labeled leaves. *Journal of Classification*, 2:7–28, 1985.

[96] J. Dean and S. Ghemawat. MapReduce: simplified data processing on large clusters. *Communications of ACM*, 51(1):107–113, 2008.

[97] P. Dehal and J. L. Boore. Two rounds of whole genome duplication in the ancestral vertebrate. *PLoS Biology*, 3, 2005.

[98] J. A. Delmerico, N. A. Byrnes, A. E. Bruno, M. D. Jones, S. M. Gallo, and V. Chaudhary. Comparing the performance of clusters, Hadoop, and Active Disks on microarray correlation computations. In *International Conference on High Performance Computing (HiPC)*, pages 378–387, 2009.

[99] R. Desper and O. Gascuel. Fast and accurate phylogeny reconstruction algorithms based on the minimum evolution principle. *Journal of Computational Biology*, 9:687–705, 2002.

[100] M. A. Dillies, A. Rau, J. Aubert, C. Hennequet-Antier, M. Jean-mougin, N. Servant, C. Keime, N.S. Marot, D. Castel, J. Estelle, G. Guernec, B. Jagla, L. Jouneau, D. Laloë, C. Le Gall, B. Schaëffer, S. Le Crom, M. Guedj, and F. Jaffrézic. A comprehensive evaluation of normalization methods for illumina high-throughput RNA sequencing data analysis. *Briefings in Bioinformatics*, 14(6):671–683, 2013.

[101] C. Ding and H. Peng. Minimum redundancy feature selection from microarray gene expression data. *Journal of Bioinformatics and Computational Biology*, 3(2):185–205, 2005.

[102] C. B. Do, M. S. P. Mahabhashyam, M. Brudno, and S. Batzoglou. ProbCons: Probabilistic consistency-based multiple sequence alignment. *Genome Research*, 15(2):330–340, 2005.

[103] R. C. Dubes and A. K. Jain. *Algorithms for Clustering Data*. Prentice Hall, NJ, 1988.

[104] R. Durbin, S. R. Eddy, A. Krogh, and G. J. Mitchison. *Biological Sequence Analysis: Probabilistic Models of Proteins and Nucleic Acids*. Cambridge University Press, 1998.

[105] S. R. Eddy. Profile hidden Markov models. *Bioinformatics*, 14:755–763, 1998.

[106] S. R. Eddy. Non-coding RNA genes and the modern RNA world. *Nature Reviews. Genetics*, 2(12):919–929, 2001.

[107] R. Edgar, M. Domrachev, and A. E. Lash. Gene expression omnibus: NCBI gene expression and hybridization array data repository. *Nucleic Acids Research*, 30(1):207–210, 2002.

[108] R. C. Edgar. MUSCLE: a multiple sequence alignment method with reduced time and space complexity. *BMC Bioinformatics*, 5, 2004.

[109] M. Egli, G. Minasov, L. Su, and A. Rich. Metal ions and flexibility in a viral RNA pseudoknot at atomic resolution. *Proceedings of National Academy of Sciences of the United States of America*, 99(7):4302–4307, 2002.

[110] E. Eisenberg and E. Y. Levanon. Human housekeeping genes are compact. *Trends In Genetics*, 19(7):362–365, 2003.

[111] R. Ekblom, T. Burke, J. Slate, and C. N. Balakrishnan. Digital gene expression analysis of the zebra finch genome. *BMC Genomics*, 11(1), 2010.

[112] Eric Ennifar and Philippe Dumas. Polymorphism of bulged-out residues in HIV-1 RNA DIS kissing complex and structure comparison with solution studies. *Journal of Molecular Biology*, 356:771–782, 2006.

[113] A. J. Enright, I. Iliopoulos, N. C. Kyrpides, and C. A. Ouzounis. Protein interaction maps for complete genomes based on gene fusion events. *Nature*, 402:86–90, 1999.

[114] T. Etzold and P. Argos. SRS: an indexing and retrieval tool for flat file data libraries. *Computer Applications in the Biosciences*, 9:49–57, 2003.

[115] J. J. Faith, B. Hayete, J. T. Thaden, I. Mogno, J. Wierzbowski, G. Cottarel, S. Kasif, J. J. Collins, and T. S. Gardner. Large-scale mapping and validation of *Escherichia coli* transcriptional regulation from a compendium of expression profiles. *PLoS Biology*, 5(1), 2007.

[116] S. Federhen. The NCBI taxonomy database. *Nucleic Acids Research*, 40:D136–D143, 2012.

[117] J. Felsenstein. *Inferring Phylogenies.* Sinauer Associates, 2003.

[118] F. Ferre, Y. Ponty, W. A. Lorenz, and P. Clote. DIAL: a Web server for the pairwise alignment of two RNA three-dimensional structures using nucleotide, dihedral angle and base-pairing similarities. *Nucleic Acids Research*, 35:W659–W668, 2007.

[119] D. Field, B. Tiwari, T. Booth, S. Houten, D. Swan, N. Bertrand, and M. Thurston. Open software for biologists: from famine to feast. *Nature Biotechnology*, 24(7):801–803, 2006.

[120] W. M. Fitch. Distinguishing homologous from analogous protein. *Systematic Zoology*, 19:99–113, 1970.

[121] S. S. Fong, A. R. Joyce, and B. ∅. Palsson. Parallel adaptive evolution cultures of *Escherichia coli* lead to convergent growth phenotypes with different gene expression states. *Genome Research*, 15(10):1365–1372, 2005.

[122] E. K. Freyhult, J. P. Bollback, and P. P. Gardner. Exploring genomic dark matter: A critical assessment of the performance of homology search methods on noncoding RNA. *Genome Research*, 17(1):117–125, 2007.

[123] N. Friedman. Inferring cellular networks using probabilistic graphical models. *Science*, (5659), 2004.

[124] N. Friedman, M. Linial, I. Nachman, and D. Pe'er. Using Bayesian networks to analyze expression data. *Journal of Computational Biology*, 7(3-4):601–620, 2000.

[125] H. Galhardas, D. Florescu, D. Shasha, E. Simon, and C. A. Saita. Declarative data cleaning: language, model, and algorithms. In *Proceedings of 27th International Conference on Very Large Data Bases*, pages 371–380, San Francisco, 2001. Morgan Kaufmann Publishers Inc.

[126] M. Y. Galperin. The molecular biology database collection: 2006 update. *Nucleic Acids Research*, 34:D3–D5, 2006.

[127] F. Gao, W. Kasprzak, V. A. Stupina, B. A. Shapiro, and A. E. Simon. A ribosome-binding, 3′ translational enhancer has a T-shaped structure and engages in a long-distance RNA-RNA interaction. *Journal of Virology*, 86(18):9828–9842, 2012.

[128] F. Gao, W. K. Kasprzak, C. Szarko, B. A. Shapiro, and A. E. Simon. The 3′ untranslated region of Pea Enation Mosaic Virus contains two T-shaped, ribosome-binding, cap-independent translation enhancers. *Journal of Virology*, 88(20), 2014.

[129] D. Gardner. Neurodatabase.org: networking the microelectrode. *Nature Neuroscience*, 7(5):486–487, 2004.

[130] D. Gardner, M. Abato, K. H. Knuth, R. DeBellis, and S. M. Erde. Dynamic publication model for neurophysiology databases. *Philosophical Transactions: Biological Sciences*, 356(1412):1229–1247, 2001.

[131] P. P. Gardner, A. Wilm, and S. Washietl. A benchmark of multiple sequence alignment programs upon structural RNAs. *Nucleic Acids Research*, 33(8):2433–2439, 2005.

[132] Michael R. Garey and David S. Johnson. *Computers and Intractability: A Guide to the Theory of NP-Completeness.* W. H. Freeman, 1979.

[133] O. Gascuel. BIONJ: an improved version of the NJ algorithm based on a simple model of sequence data. *Molecular Biology and Evolution*, 14:685–695, 1997.

[134] E. Gasteiger, E. Jung, and A. Bairoch. SWISS-PROT: connecting biomolecular knowledge via a protein database. *Current Issues in Molecular Biology*, 3:47–55, 2001.

[135] P. Geurts, D. Ernst, and L. Wehenkel. Extremely randomized trees. *Machine Learning*, 63(1):3–42, 2006.

[136] S. Ghemawat, H. Gobioff, and S. Leung. The Google file system. *SIGOPS Operating Systems Review*, 37(5):29–43, 2003.

[137] S. D. Gilbert, C. D. Stoddard, S. J. Wise, and R. T. Batey. Thermodynamic and kinetic characterization of ligand binding to the purine riboswitch aptamer domain. *Journal of Molecular Biology*, 359:754–768, 2006.

[138] Z. Gillani, M. S. H. Akash, M. D. M. Rahaman, and M. Chen. CompareSVM: supervised, support vector machine (SVM) inference of gene regularity networks. *BMC Bioinformatics*, 15(1), 2014.

[139] J. Goecks, J. Taylor, A. Nekrutenko, E. Afgan, G. Ananda, D. Baker, D. Blankenberg, R. Chakrabarty, N. Coraor, J. Goecks, G. Von Kuster, R. Lazarus, K. Li, A. Nekrutenko, J. Taylor, and

K. Vincent. Galaxy: a comprehensive approach for supporting accessible, reproducible, and transparent computational research in the life sciences. *Genome Biology*, 11(8), 2010.

[140] A. Goffeau, B. G. Barrell, H. Bussey, R. W. Davis, B. Dujon, H. Feldmann, F. Galibert, J. D. Hoheisel, C. Jacq, M. Johnston, E. J. Louis, H. W. Mewes, Y. Murakami, P. Philippsen, H. Tettelin, and S. G. Oliver. Life with 6000 genes. *Science*, (5287), 1996.

[141] A. Greenfield, R. Bonneau, A. Madar, and H. Ostrer. DREAM4: Combining genetic and dynamic information to identify biological networks and dynamical models. *PLoS One*, 5(10), 2010.

[142] S. Griffiths-Jones, A. Bateman, M. Marshall, A. Khanna, and S. R. Eddy. Rfam: an RNA family database. *Nucleic Acids Research*, 31:439–441, 2003.

[143] S. Griffiths-Jones, A. Khanna, S. R. Eddy, S. Moxon, M. Marshall, and A. Bateman. Rfam: Annotating non-coding RNAs in complete genomes. *Nucleic Acids Research*, 33:D121–D124, 2005.

[144] K. Grzeskowiak, K. Yanagi, G. G. Prive, and R. E. Dickerson. The structure of B-helical C-G-A-T-C-G-A-T-C-G, and comparison with C-C-A-A-C-G-T-T-G-G: the effect of base pair reversal. *Journal of Biological Chemistry*, 266:8861–8883, 1991.

[145] S. Guindon and O. Gascuel. A simple, fast and accurate method to estimate large phylogenies by maximum-likelihood. *Systems Biology*, 52:696–704, 2003.

[146] T. Gunarathne, B. Zhang, T. Wu, and J. Qiu. Scalable parallel computing on clouds using Twister4Azure iterative MapReduce. *Future Generation Computer Systems*, 29(4):1035–1048, 2013.

[147] J. Gurtowski, M. C. Schatz, and B. Langmead. Genotyping in the Cloud with Crossbow. *Current Protocols in Bioinformatics*, 2012.

[148] D. Gusfield. *Algorithms on Strings, Trees, and Sequences : Computer Science and Computational Biology*. Cambridge University Press, 1997.

[149] I. Guyon and A. Elisseeff. An introduction to variable and feature selection. *Journal of Machine Learning Research*, 3(7/8):1157–1182, 2003.

[150] B. J. Haas, J. R. Wortman, C. M. Ronning, L. I. Hannick, R. K. Smith Jr., R. Maiti, A. P. Chan, Y. Chunhui, M. Farzad,

W. Dongying, O. White, and C. D. Town. Complete reannotation of the *Arabidopsis* genome: methods, tools, protocols and the final release. *BMC Biology*, 3:1–19, 2005.

[151] F. T. Haddadin and S. W. Harcum. Transcriptome profiles for high-cell-density recombinant and wild-type *Escherichia coli*. *Biotechnology And Bioengineering*, 90(2):127–153, 2005.

[152] A. Y. Halevy. Answering queries using views: a survey. *Very Large Database Journal*, 10(4):270–294, 2001.

[153] A. Y. Halevy. Data integration: a status report. In *Proceedings of Tenth Conference on Database Systems for Business, Technology and the Web*, pages 24–29, Germany, 2003.

[154] A. Y. Halevy, A. G. Ives, P. Mork, and I. Tatarinov. Piazza: data management infrastructure for semantic Web applications. In *Proceedings of Twelfth International World Wide Web Conference*, pages 556–567, Budapest, 2003. ACM.

[155] B. Han, B. Dost, V. Bafna, and S. Zhang. Structural alignment of pseudoknotted RNA. *Journal of Computational Biology*, 15(5):489–504, 2008.

[156] J. Han and M. Kamber. *Data Mining: Concepts and Techniques*. Morgan Kaufmann, San Francisco, 2006.

[157] J. Han, M. Kamber, and J. Pei. *Data Mining: Concepts and Techniques. 3rd ed.* Morgan Kaufmann, 2011.

[158] D. J. Hand, H. Mannila, and P. Smyth. *Principles of Data Mining*. MIT Press, 2001.

[159] B. Hayete, T. S. Gardner, and J. J. Collins. Size matters: network inference tackles the genome scale. *Molecular Systems Biology*, 3, 2007.

[160] G. He, A. Steppi, J. Laborde, A. Srivastava, P. Zhao, and J. Zhang. RASS: a Web server for RNA alignment in the joint sequence-structure space. *Nucleic Acids Research*, 42:W377–W381, 2014.

[161] J. Hein, T. Jiang, L. Wang, and K. Zhang. On the complexity of comparing evolutionary trees. *Discrete Applied Mathematics*, 71:153–169, 1996.

[162] K. G. Herbert, J. Westbrook, and J. T. L. Wang. Data integration in biological databases. In *Proceedings of Atlantic Symposium*

on Computational Biology and Genome Information Systems &
Technology, Durham, 2003.

[163] M. A. Hernandez and S. J. Stolfo. The merge/purge problem for
large databases. In *Proceedings of ACM SIGMOD International
Conference on Management of Data*, pages 127–138, 1995.

[164] M. A. Hernandez and S. J. Stolfo. Real-world data is dirty: data
cleansing and the merge/purge problem. *Data Mining and Knowledge Discovery*, 2:9–37, 1998.

[165] A. Herraez. Biomolecules in the computer: Jmol to the rescue.
Biochemical Education, 34:255–261, 2006.

[166] C. Hertz-Fowler, C. S. Peacock, V. Wood, M. Aslett, A. Ker-
hornou, P. Mooney, A. Tivey, M. Berriman, N. Hall, K. Rutherford, J. Parkhill, A. C. Ivens, M. A. Rajandream, and B. Barrell.
GeneDB: a resource for prokaryotic and eukaryotic organisms. *Nucleic Acids Research*, 32:D339–D343, 2004.

[167] M. Höchsmann, T. Töller, R. Giegerich, and S. Kurtz. Local similarity in RNA secondary structures. In *Proceedings of IEEE Computer Society Bioinformatics Conference*, volume 2, pages 159–168,
2003.

[168] M. Höchsmann, B. Voss, and R. Giegerich. Pure multiple RNA
secondary structure alignments: A progressive profile approach.
*IEEE/ACM Transactions on Computational Biology and Bioin-
formatics*, 1(1):53–62, 2004.

[169] I. L. Hofacker. Vienna RNA secondary structure server. *Nucleic
Acids Research*, 31(13):3429–3431, 2003.

[170] I. L. Hofacker, P. F. Stadler, M. Tacker, P. Schuster, W. Fontana,
and L. S. Bonhoeffer. Fast folding and comparison of RNA secondary structures. *Monatshefte für Chemie Chemical Monthly*,
125(2):167–188, 1989.

[171] D. Hoksza and D. Svozil. Efficient RNA pairwise structure comparison by SETTER method. *Bioinformatics*, 28(14):1858–1864,
2012.

[172] J. Hu, H. Ge, M. Newman, and K. Liu. OSA: a fast and accurate
alignment tool for RNA-seq. *Bioinformatics*, 28(14):1933–1934,
2012.

[173] Z. Hu, P. J. Killion, and V. R. Iyer. Genetic reconstruction of a functional transcriptional regulatory network. *Nature Genetics*, 39(5):683–687, 2007.

[174] Z. Z. Hu, I. Mani, V. Hermoso, H. Liu, and C. H Wu. iProLINK: an integrated protein resource for literature mining. *Computational Biology and Chemistry*, 28:409–416, 2004.

[175] H. Huang, W. C. Barker, Y. Chen, and C. H. Wu. iProClass: an integrated database of protein family, function and structure information. *Nucleic Acids Research*, 31:390–392, 2003.

[176] Z. Huang, Y. Wu, J. Robertson, L. Feng, R. Malmberg, and L. Cai. Fast and accurate search for non-coding RNA pseudoknot structures in genomes. *Bioinformatics*, 24(20):2281–2287, 2008.

[177] T. Hubbard, D. Barker, L. Clark, T. Cox, J. Cuff, V. Curwen, T. Down, R. Durbin, E. Eyras, J. Gilbert, R. Pettett, M. Pocock, S. Potter, S. Searle, J. Smith, W. Spooner, J. Stalker, M. Clamp, E. Birney, G. Cameron, Y. Chen, M. Hammond, L. Huminiecki, A. Kasprzyk, H. Lehvaslaiho, P. Lijnzaad, C. Melsopp, E. Mongin, A. Rust, E. Schmidt, G. Slater, A. Stabenau, E. Stupka, A. Ureta-Vidal, and I. Vastrik. The Ensembl genome database project. *Nucleic Acids Research*, 30(1):38–41, 2002.

[178] A. M. Huerta, H. Salgado, D. Thieffry, and J. Collado-Vides. RegulonDB: a database on transcriptional regulation in *Escherichia coli*. *Nucleic Acids Research*, 26(1):55–59, 1998.

[179] N. Hulo, A. Bairoch, V. Bulliard, L. Cerutti, E. DeCastro, P. S. Langendijk-Genevaux, M. Pagni, and C. J. Sigrist. The PROSITE database. *Nucleic Acids Research*, 34(1):D227—D230, 2006.

[180] D. G. Hurley, J. Cursons, Y. K. Wang, D. M. Budden, C. G. Print, and E. J. Crampin. NAIL, a software toolset for inferring, analyzing and visualizing regulatory networks. *Bioinformatics (Oxford, England)*, 2014.

[181] V. A. Huynh-Thu, A. Irrthum, L. Wehenkel, and P. Geurts. Inferring regulatory networks from expression data using tree-based methods. *PLoS One*, 5(9):1–10, 09 2010.

[182] V. A. Huynh-Thu and G. Sanguinetti. Combining tree-based and dynamical systems for the inference of gene regulatory networks. *Bioinformatics*, 31(10):1614–1622, 2014.

[183] Z. Ives, A. Y. Levy, J. Madhavan, R. Pottinger, S. Saroiu, I. Tatari-nov, S Betzler, Q. Chen, E. Jaslikowska, J. Su, and W. T. T. Ye-ung. Self-organizing data sharing communities with SAGRES. In *Proceedings of 2000 ACM SIGMOD International Conference on Management of Data*, Dallas, 2000. ACM.

[184] M. Jarke and Y. Vassiliou. Data warehouse quality: a review of the DWQ project. In *Proceedings of 2nd Conference on Information Quality*. Massachusetts Institute of Technology, 1997.

[185] T. Jiang, G. Lin, B. Ma, and K. Zhang. A general edit distance between RNA structures. *Journal of Computational Biology*, 9(2):371–388, 2002.

[186] T. Joachims. Making large scale SVM learning practical. Technical report, Universität Dortmund, 1999.

[187] F. Jourdan, R. Breitling, M. P. Barrett, and D. Gilbert. MetaNet-ter: Inference and visualization of high-resolution metabolomic networks. *Bioinformatics*, 24(1):143–145, 2008.

[188] L. Jourdren, M. Bernard, S. Le Crom, and M. A. Dillies. Eoulsan: A cloud computing-based framework facilitating high throughput sequencing analyses. *Bioinformatics*, 28(11):1542–1543, 2012.

[189] M. Kanehisa. Pathway databases and higher order function. *Advances in Protein Chemistry*, 54:381–408, 2000.

[190] M. Kanehisa. *Post-Genome Informatics*. Oxford University Press, UK, 2000.

[191] M. Kanehisa and S. Goto. KEGG: Kyoto Encyclopedia of Genes and Genomes. *Nucleic Acids Research*, 28:27–30, 2000.

[192] M. Kanehisa, S. Goto, S. Kawashima, and A. Nakaya. The KEGG databases at GenomeNet. *Nucleic Acids Research*, 30:42–46, 2002.

[193] S. Kannan, E. Lawler, and T. Warnow. Determining the evolution-ary tree. In *Proceedings of 1st Annual ACM-SIAM Symposium on Discrete Algorithms*, pages 475–484, 1990.

[194] S. Kannan, T. Warnow, and S. Yooseph. Computing the local consensus of trees. In *Proceedings of 6th Annual ACM-SIAM Symposium on Discrete Algorithms*, pages 68–77, 1995.

[195] M. Y. Kao, T. W. Lam, T. M. Przytycka, W. K. Sung, and H. F. Ting. General techniques for comparing unrooted evolutionary

trees. In *Proceedings of 29th Annual ACM Symposium on Theory of Computing*, pages 54–65, 1997.

[196] P. D. Karp, C. A. Ouzounis, C. Moore-Kochlacs, L. Goldovsky, P. Kaipa, D. Ahren, S. Tsoka, N. Darzentas, V. Kunin, and N. Lopez-Bigas. Expansion of the BioCyc collection of pathway/genome databases to 160 genomes. *Nucleic Acids Research*, 19:6083–6089, 2005.

[197] L. A. Kavanaugh and F. S. Dietrich. Non-coding RNA prediction and verification in *Saccharomyces cerevisiae*. *PLoS Genetics*, 5(1), 2009.

[198] J. Kennedy and R. Eberhart. Particle swarm optimization. *Proceedings of International Conference on Neural Networks*, (4), 1995.

[199] W. J. Kent, C. W. Sugnet, T. S. Furey, K. M. Roskin, T. H. Pringle, A. M. Zahler, and D. Haussler. The human genome browser at UCSC. *Genome Research*, 12(6):996–1006, 2002.

[200] I. M. Keseler, A. Mackie, M. Peralta-Gil, A. Santos-Zavaleta, S. Gama-Castro, C. Bonavides-Martínez, C. Fulcher, A. M. Huerta, A. Kothari, M. Krummenacker, M. Latendresse, L. Muñiz Rascado, Q. Ong, S. Paley, I. Schröder, A. G. Shearer, P. Subhraveti, M. Travers, D. Weerasinghe, V. Weiss, J. Collado-Vides, R. P. Gunsalus, I. Paulsen, and P. D. Karp. EcoCyc: fusing model organism databases with systems biology. *Nucleic Acids Research*, 41:D605–D612, 2013.

[201] M. Khaladkar, J. Liu, D. Wen, J. T. L. Wang, and B. Tian. Mining small RNA structure elements in untranslated regions of human and mouse mRNAs using structure-based alignment. *BMC Genomics*, 9, 2008.

[202] M. Khaladkar, V. Patel, V. Bellofatto, J. Wilusz, and J. T. L. Wang. Detecting conserved secondary structures in RNA molecules using constrained structural alignment. *Computational Biology and Chemistry*, 32(4):264–272, 2008.

[203] M. Khaladkar, J. T. L. Wang, V. Bellofatto, B. Tian, and B. A. Shapiro. RADAR: A Web server for RNA data analysis and research. *Nucleic Acids Research*, 35:W300–W304, 2007.

[204] F. Kieken, F. Paquet, F. Brulé, J. Paoletti, and G. Lancelot. A new NMR solution structure of the SL1 HIV-1Lai loop-loop dimer. *Nucleic Acids Research*, 34(1):343–352, 2006.

[205] J. Kim, A. E. Walter, and D. H. Turner. Thermodynamics of coaxially stacked helixes with GA and CC mismatches. *Biochemistry*, 35:13753–13761, 1996.

[206] S. H. Kim, J. L. Sussman, F. L. Suddath, G. J. Quigley, A. McPherson, A. H. J. Wang, N. C. Seeman, and A. Rich. The general structure of transfer RNA molecules. *Proceedings of National Academy of Sciences of the United States of America*, (12):4970–4974, 1974.

[207] T. Kin, K. Yamada, G. Terai, H. Okida, Y. Yoshinari, Y. Ono, A. Kojima, Y. Kimura, T. Komori, and K. Asai. fRNAdb: a platform for mining/annotating functional RNA candidates from non-coding RNA sequences. *Nucleic Acids Research*, 35:D145–D148, 2007.

[208] C. Kozanitis and D. A. Patterson. GenAp: A distributed SQL interface for genomic data. *BMC Bioinformatics*, 2016.

[209] S. Krishnaprasad. Uses and abuses of Amdahl's law. *Journal of Computing Sciences in Colleges*, 17(2):288–293, 2001.

[210] E. Kubicka, G. Kubicki, and F. R. McMorris. An algorithm to find agreement subtrees. *Journal of Classification*, 12(1):91–99, 1995.

[211] A. Lacroix. The biological data integration system. In *Proceedings of Fifth ACM International Workshop on Web Information and Data Management*, pages 45–49, New Orleans, 2003.

[212] Z. Lacroix. Biological data integration: wrapping data and tools. *IEEE Transactions on Information Technology in Biomedicine*, 6:123–128, 2002.

[213] Z. Lacroix and T. Critchlow, editors. *Bioinformatics: Managing Scientific Data*. Morgan Kaufmann, Burlington, MA, 2003.

[214] H. Lähdesmäki, O. Yli-Harja, and I. Shmulevich. On learning gene regulatory networks under the Boolean network model. *Machine Learning*, 52(1-2):147–167, 2003.

[215] C. Laing, D. Wen, J. T. L. Wang, and T. Schlick. Predicting coaxial helical stacking in RNA junctions. *Nucleic Acids Research*, 40:487–498, 2012.

[216] T. Lam, W. Sung, and H. Ting. Computing the unrooted maximum agreement subtree in subquadratic time. In *Proceedings of 5th Scandinavian Workshop on Algorithm Theory*, pages 124–135, 1996.

[217] E. S. Lander, L. M. Linton, B. Birren, C. Nusbaum, M. C. Zody, J. Baldwin, K. Devon, K. Dewar, M. Doyle, W. Fitzhugh, R. Funke, D. Gage, K. Harris, A. Heaford, J. Howland, L. Kann, J. Lehoczky, R. Levine, P. McEwan, K. McKernan, J. Meldrim, J. P. Mesirov, C. Miranda, W. Morris, J. Naylor, C. Raymond, M. Rosetti, R. Santos, A. Sheridan, C. Sougnez, N. Stange-Thomann, N. Stojanovic, A. Subramanian, D. Wyman, S. Batzoglou, D. G. Brown, J. Galagan, V. J. Pollara, J. Rogers, J. Sulston, R. Ainscough, S. Beck, D. Bentley, J. Burton, C. Clee, N. Carter, A. Coulson, R. Deadman, P. Deloukas, A. Dunham, I. Dunham, R. Durbin, L. French, D. Grafham, S. Gregory, T. Hubbard, S. Humphray, A. Hunt, M. Jones, C. Lloyd, A. McMurray, L. Matthews, S. Mercer, S. Milne, J. C. Mullikin, A. Mungall, R. Plumb, M. Ross, R. Shownkeen, S. Sims, A. Bateman, M. Clamp, J. G. R. Gilbert, R. H. Waterston, R. K. Wilson, L. W. Hillier, J. D. McPherson, M. A. Marra, E. R. Mardis, L. A. Fulton, A. T. Chinwalla, K. H. Pepin, W. R. Gish, S. L. Chissoe, M. C. Wendl, K. D. Delehaunty, T. L. Miner, A. Delehaunty, J. B. Kramer, L. L. Cook, R. S. Fulton, D. L. Johnson, P. J. Minx, S. W. Clifton, I. Korf, J. Wallis, S. P. Yang, T. Hawkins, E. Branscomb, P. Predki, P. Richardson, S. Wenning, T. Slezak, N. Doggett, J. F. Cheng, A. Olsen, S. Lucas, C. Elkin, E. Uberbacher, M. Frazier, R. A. Gibbs, D. M. Muzny, S. E. Scherer, J. B. Bouck, E. J. Sodergren, K. C. Worley, C. M. Rives, J. H. Gorrell, M. L. Metzker, S. L. Naylor, R. S. Kucherlapati, D. L. Nelson, G. M. Weinstock, Y. Sakaki, A. Fujiyama, M. Hattori, T. Yada, A. Toyoda, T. Itoh, C. Kawagoe, H. Watanabe, Y. Totoki, T. Taylor, J. Weissenbach, R. Heilig, W. Saurin, F. Artiguenave, P. Brottier, T. Bruls, E. Pelletier, C. Robert, P. Wincker, D. R. Smith, L. Doucette-Stamm, M. Rubenfield, K. Weinstock, M. L. Hong, J. Dubois, A. Rosenthal, M. Platzer, G. Nyakatura, S. Taudien, A. Rump, H. Yang, J. Yu, J. Wang, G. Huang, J. Gu, L. Hood, L. Rowen, A. Madan, S. Qin, R. W. Davis, N. A. Federspiel, A. P. Abola, M. J. Proctor, R. M. Myers, J. Schmutz, M. Dickson, J. Grimwood, D. R. Cox, M. V. Olson, R. Kaul, C. Raymond, N. Shimizu, K. Kawasaki, S. Minoshima, G. A. Evans, M. Athanasiou, R. Schultz, B. A. Roe, F. Chen, H. Pan, J. Ramser, H. Lehrach, R. Reinhardt, W. R. McCombie, M. De La Bastide, N. Dedhia, H. Blöcker, K. Hornischer, G. Nordsiek, R. Agarwala, L. Aravind, H. C. Chen, D. Church, W. Jang, P. Kitts, E. V. Koonin, G. Schuler, D. Thierry-Mieg, J. Thierry-Mieg, L. Wagner, Y. I. Wolf, J. A. Bailey, E. E. Eichler, E. Birney, L. Cerutti, H. Hermjakob, A. Kaspryzk, N. Mulder, G. Slater, E. Stupka, P. Bork, T. Doerks, R. R. Copley,

J. Schultz, C. B. Burge, R. F. Yeh, S. R. Eddy, L. S. Johnson, T. A. Jones, T. S. Furey, C. Harmon, D. Kulp, R. Wheeler, A. Williams, Y. Hayashizaki, D. Haussler, K. Hokamp, A. McLysaght, K. H. Wolfe, S. Kasif, T. Mikkelsen, J. Szustakowki, S. Kennedy, W. J. Kent, D. Lancet, T. M. Lowe, J. V. Moran, C. P. Ponting, A. F. A. Smit, F. Collins, M. S. Guyer, J. Peterson, A. Felsenfeld, K. A. Wetterstrand, A. Patrinos, and M. J. Morgan. Initial sequencing and analysis of the human genome. *Nature*, 409(6822):860–921, 2001.

[218] B. Langmead. Aligning short sequencing reads with Bowtie. *Current Protocols in Bioinformatics*, (32), 2010.

[219] B. Langmead, K. D. Hansen, and J. T. Leek. Cloud-scale RNA-sequencing differential expression analysis with Myrna. *Genome Biology*, 11(8), 2010.

[220] B. Langmead and S. L. Salzberg. Fast gapped-read alignment with Bowtie 2. *Nature Methods*, 9(4):357–359, 2012.

[221] M. E. Laubacher and S. E. Ades. The Rcs phosphorelay is a cell envelope stress response activated by peptidoglycan stress and contributes to intrinsic antibiotic resistance. *Journal of Bacteriology*, 190(6):2065–2074, 2008.

[222] R. Lavery and H. Sklenar. The definition of generalized helicoidal parameters and of axis curvature for irregular nucleic acids. *Journal of Biomolecular Structure and Dynamics*, 6(1):63–91, 1988.

[223] I. Lebars, T. Richard, C. Di Primo, and J. J. Toulm. NMR structure of a kissing complex formed between the TAR RNA element of HIV-1 and a LNA-modified aptamer. *Nucleic Acids Research*, 35(18):6103–6114, 2007.

[224] Y. Lee, Y. Hsiao, and W. Hwang. Designing a parallel evolutionary algorithm for inferring gene networks on the cloud computing environment. *BMC Systems Biology*, 8, 2014.

[225] M. Lenzerini. Data integration: a theoretical perspective. In *Proceedings of ACM SIGACT-SIGMOD-SIGART Symposium on Principles of Database Systems*, pages 233–246, 2002.

[226] S. Leo, F. Santoni, and G. Zanetti. Biodoop: Bioinformatics on Hadoop. In *ICPP Workshops*, pages 415–422, 2009.

[227] N. B. Leontis, A. Lescoute, and E. Westhof. The building blocks and motifs of RNA architecture. *Current Opinions in Structural Biology*, 16:279–287, 2006.

[228] G. Leroy and H. Chen. Meeting medical terminology needs: the ontology-enhanced medical concept mapper. *IEEE Transactions on Information Technology in Biomedicine*, 5(4):261–270, 2001.

[229] G. Leroy and H. Chen. Genescene: biomedical text and data mining. In *Proceedings of Joint Conference on Digital Libraries*, pages 116–118, 2003.

[230] A. Lescoute and E. Westhof. Topology of three-way junctions in folded RNAs. *RNA*, 12(1):83–93, 2006.

[231] A. Lesk. *Database Annotation in Molecular Biology*. John Wiley & Sons, 2005.

[232] S Lewis, A Csordas, S Killcoyne, H Hermjakob, MR Hoopmann, RL Moritz, EW Deutsch, and J Boyle. Hydra: a scalable proteomic search engine which utilizes the Hadoop distributed computing framework. *BMC Bioinformatics*, 13, 2012.

[233] A. Liaw and M. Wiener. Classification and regression by random-Forest. *R. News*, 2:18–22, 2002.

[234] J. S. Lim, B. S. Choi, J. S. Lee, C. Shin, T. J. Yang, J. S. Rhee, J. S. Lee, and I. Y. Choi. Survey of the applications of NGS to whole-genome sequencing and expression profiling. *Genomics & Informatics*, 10(1):1–8, 2012.

[235] J. M. Lingeman and D. Shasha. *Network Inference in Molecular Biology: A Hands-on Framework*. Springer, 2012.

[236] C. E. Lipscomb. Medical subject headings (MeSH). *Bulletin of Medical Library Association*, 88(3):265–266, 2000.

[237] C. Liu, B. Bai, G. Skogerbø, L. Cai, W. Deng, Y. Zhang, D. Bu, Y. Zhao, and R. Chen. NONCODE: an integrated knowledge database of non-coding RNAs. *Nucleic Acids Research*, 33:D112–D115, 2005.

[238] J. Liu, J. T. L. Wang, J. Hu, and B. Tian. A method for aligning RNA secondary structures and its application to RNA motif detection. *BMC Bioinformatics*, 6, 2005.

[239] A. E. Loraine, I. C. Blakley, S. Jagadeesan, J. Harper, G. Miller, and N. Firon. Analysis and visualization of RNA-seq expression data using RStudio, Bioconductor, and Integrated Genome Browser. *Methods in Molecular Biology*, 1284:481–501, 2015.

[240] W. L. Low, M. L. Lee, and T. W. Ling. A knowledge-based approach for duplicate elimination in data cleaning. *Information Systems*, 26(8):585–606, 2001.

[241] R. Luo, T. Wong, J. Zhu, C. M. Liu, E. Wu, L. K. Lee, D. W. Cheung, H. F. Ting, S. M. Yiu, T. W. Lam, X. Zhu, S. Peng, H. Lin, W. Zhu, C. Yu, Y. Li, and R. Li. SOAP3-dp: fast, accurate and sensitive GPU-based short read aligner. *PLoS One*, 8(5), 2013.

[242] K. D. MacIsaac, T Wang, D. B. Gordon, D. K. Gifford, G. D. Stormo, and E. Fraenkel. An improved map of conserved regulatory sites for *Saccharomyces cerevisiae*. *BMC Bioinformatics*, (1), 2006.

[243] A. Madar, A. Greenfield, H. Ostrer, E. Vanden-Eijnden, and R. Bonneau. The Inferelator 2.0: a scalable framework for reconstruction of dynamic regulatory network models. *Annual International Conference of the IEEE Engineering in Medicine and Biology Society.*, pages 5448–5451, 2009.

[244] S. R. Maetschke, P. B. Madhamshettiwar, M. J. Davis, and M. A. Ragan. Supervised, semi-supervised and unsupervised inference of gene regulatory networks. *Briefings in Bioinformatics*, 15(2):195–211, 2014.

[245] Y. Maki, T. Ueda, M. Okamoto, N. Uematsu, K. Inamura, K. Uchida, Y. Takahashi, and Y. Eguchi. Inference of genetic network using the expression profile time course data of mouse P19 cells. *Genome Informatics*, 13:382–383, 2002.

[246] D. Marbach, R. Küffner, M. Kellis, B. Holmes, J. C. Costello, N. M. Vega, D. M. Camacho, K. R. Allison, J. J. Collins, N. Vega, T. Petri, L. Windhager, R. Zimmer, R. J. Prill, G. Stolovitzky, A. Aderhold, R. Bonneau, F. Dondelinger, D. Husmeier, A. Madar, C. S. Poultney, A. Greenfield, S. Mani, Y. Chen, F. Cordero, R. Esposito, A. Visconti, M. Crane, H. J. Ruskin, A. Sîrbu, M. Drton, R. Foygel, S. Rezny, A. De La Fuente, V. De Leo, A. Pinna, N. Soranzo, J. Gertheiss, T. Hothorn, P. Geurts, V. A. Huynh-Thu, A. Irrthum, L. Wehenkel, M. Grzegorczyk, A. C. Haury, F. Mordelet, P. Vera-Licona, J. P. Vert, G. Karlebach, R. Shamir, S. Lèbre, H. Ostrer, Z. Ouyang, M. Song, H. Wang, Y. Zhang, R. Pandya, Y. Saeys, and A. Statnikov. Wisdom of crowds for robust gene network inference. *Nature Methods*, 9(8):796–804, 2012.

[247] D. Marbach, R. J. Prill, T. Schaffter, C. Mattiussi, D. Floreano, G. Stolovitzky, and C. R. Cantor. Revealing strengths and weak-

nesses of methods for gene network inference. *Proceedings of National Academy of Sciences of the United States of America*, (14), 2010.

[248] D Marbach, S Roy, F Ay, PE Meyer, R Candeias, T Kahveci, CA Bristow, and M Kellis. Predictive regulatory models in *Drosophila melanogaster* by integrative inference of transcriptional networks. *Genome Research*, 22:1334–1349, 2012.

[249] D. Marbach, T. Schaffter, C. Mattiussi, and D. Floreano. Generating realistic in silico gene networks for performance assessment of reverse engineering methods. *Journal of Computational Biology*, 16(2):229–239, 2009.

[250] A. A. Margolin, A. Califano, I. Nemenman, C. Wiggins, K. Basso, R. D. Favera, and G. Stolovitzky. ARACNE: An algorithm for the reconstruction of gene regulatory networks in a mammalian cellular context. *BMC Bioinformatics*, 7, 2006.

[251] J. C. Marioni, M. Stephens, Y. Gilad, C. E. Mason, and S. M. Mane. RNA-seq: an assessment of technical reproducibility and comparison with gene expression arrays. *Genome Research*, 18(9):1509–1517, 2008.

[252] D. H. Mathews and D. H. Turner. Dynalign: an algorithm for finding the secondary structure common to two RNA sequences. *Journal of Molecular Biology*, 317(2):191–203, 2002.

[253] A. M. Matsunaga, M. O. Tsugawa, and J. A. B. Fortes. Cloud-BLAST: Combining MapReduce and virtualization on distributed resources for bioinformatics applications. In *eScience*, pages 222–229, 2008.

[254] J. S. Mattick and I. V. Makunin. Non-coding RNA. *Human Molecular Genetics*, 15:R17–R29, 2006.

[255] J. S. McCaskill. The equilibrium partition function and base pair binding probabilities for RNA secondary structure. *Biopolymers*, 29(6-7):1105–1119, 1990.

[256] J. C. McCormack, X. Yuan, R. E. Zamora, A. E. Simon, Y. G. Yingling, B. A. Shapiro, and W. Kasprzak. Structural domains within the untranslated region of Turnip crinkle virus. *Journal of Virology*, 82(17):8706–8720, 2008.

[257] T. Meinel, A. Krause, H. Luz, M. Vingron, and E. Staub. The SYSTERS protein family database in 2005. *Nucleic Acids Research*, 33:D226–D229, 2005.

[258] D. W. Meinke, J. M. Cherry, C. Dean, S. D. Rounsley, and M. Koornneef. *Arabidopsis thaliana*: a model plant for genome analysis.

[259] I. M. Meyer and I. Miklos. SimulFold: simultaneously inferring RNA structures including pseudoknots, alignments, and trees using a Bayesian MCMC framework. *PLoS Computational Biology*, 3(8), 2007.

[260] P. E. Meyer, K. Kontos, F. Lafitte, and G. Bontempi. Information-theoretic inference of large transcriptional regulatory networks. *EURASIP Journal on Bioinformatics and Systems Biology*, 2007.

[261] P. E. Meyer, F. Lafitte, and G. Bontempi. *minet*: an R/bioconductor package for inferring large transcriptional networks using mutual information. *BMC Bioinformatics*, 9, 2008.

[262] T. M. Mitchell. *Machine Learning*. McGraw-Hill, 1997.

[263] S. Miyazawa. A reliable sequence alignment method based on probabilities of residue correspondences. *Protein Engineering*, 8(10):999–1009, 1995.

[264] M. Möhl, S. Will, and R. Backofen. Lifting prediction to alignment of RNA pseudoknots. *Journal of Computational Biology*, 17(3):429–442, 2010.

[265] E. J. Molinelli, A. Korkut, W. Wang, M. L. Miller, N. P. Gauthier, X. Jing, P. Kaushik, Q. He, C. Sander, G. Mills, D. B. Solit, C. A. Pratilas, M. Weigt, A. Braunstein, A. Pagnani, and R. Zecchina. Perturbation biology: inferring signaling networks in cellular systems. *PLoS Computational Biology*, 9(12), 2013.

[266] F. Mordelet and J.-. P. Vert. SIRENE: supervised inference of regulatory networks. *Bioinformatics*, 24(16):i76–i82, 2008.

[267] Z. Mousavian, K. Kavousi, and A. Masoudi-Nejad. Review: Information theory in systems biology. Part I: Gene regulatory and metabolic networks. *Seminars in Cell and Developmental Biology*, 2015.

[268] A. Mujeeb, T. G. Parslow, A. Zarrinpar, C. Das, and T. L. James. NMR structure of the mature dimer initiation complex of HIV-1 genomic RNA. *FEBS Letters*, 458:387–392, 1999.

[269] N. J. Mulder, R. Apweiler, D. Binns, P. Bradley, U. Das, W. Fleischmann, N. Harte, A. Kanapin, M. Krestyaninova, D. Lonsdale, R. Lopez, J. Maslen, J. McDowall, A. Mitchell, S. Orchard, E. Quevillon, V. Silventoinen, R. Vaughan, A. Bateman, R. Durbin, D. J. Studholme, T. K. Attwood, A. Bairoch, N. Hulo, C. J. A. Sigrist, P. Bork, I. Letunic, P. Bucher, L. Cerutti, M. Pagni, R. Copley, E. Courcelle, D. Kahn, J. Gough, D. Haft, J. Selengut, M. Madera, A. N. Nikolskaya, C. H. Wu, and C. P. Ponting. InterPro, progress and status in 2005. *Nucleic Acids Research*, 33:D201–D205, 2005.

[270] E. P. Nawrocki and S. R. Eddy. Infernal 1.1: 100-fold faster RNA homology searches. *Bioinformatics*, 29(22):2933–2935, 2013.

[271] E. P. Nawrocki, D. L. Kolbe, and S. R. Eddy. Infernal 1.0: inference of RNA alignments. *Bioinformatics*, 25:1335–1337, 2009.

[272] G. Neudert and G. Klebe. FCONV: format conversion, manipulation and feature computation of molecular data. *Bioinformatics*, 27(7):1021–1022, 2011.

[273] T. Nguyen, W. Shi, and D. Ruden. CloudAligner: A fast and full-featured MapReduce-based tool for sequence mapping. *BMC Research Notes*, 4(1):171–177, 2011.

[274] Z. Nie and S. Kambhampati. A frequency-based approach for mining coverage statistics in data integration. In *Proceedings of 20th International Conference on Data Engineering*, Boston, 2004.

[275] Z. Nie, S. Kambhampati, and T. Hernandez. BibFinder/StatMiner: effectively mining and using coverage and overlap statistics in data integration. In *Proceedings of 29th International Conference on Very Large Data Bases*, pages 1097–1100, Berlin, Germany, 2003.

[276] P. L. Nixon, A. Rangan, Y. G. Kim, A. Rich, D. W. Hoffman, M. Hennig, and D. P. Giedroc. Solution structure of a luteoviral P1-P2 frameshifting mRNA pseudoknot. *Journal of Molecular Biology*, 322:621–633, 2002.

[277] P. A. Nuin, Z. Wang, and E. R. Tillier. The accuracy of several multiple sequence alignment programs for proteins. *BMC Bioinformatics*, 7, 2006.

[278] A. Ocone, G. Sanguinetti, and A. J. Millar. Hybrid regulatory models: a statistically tractable approach to model regulatory network dynamics. *Bioinformatics*, 29(7):910–916, 2013.

[279] C. O'Donovan, M. J. Martin, A. Gattiker, E. Gasteiger, A. Bairoch, and R. Apweiler. High-quality protein knowledge resource: SWISSPROT and TrEMBL. *Briefings in Bioinformatics*, 3:275–284, 2002.

[280] K. Okubo, H. Sugawara, T. Gojobori, and Y. Tateno. DDBJ in preparation for overview of research activities behind data submissions. *Nucleic Acids Research*, 34(1):D6–D9, 2006.

[281] D. A. Orlando, C. Y. Lin, A. Bernard, J. Y. Wang, J. E. S. Socolar, E. S. Iversen, A. J. Hartemink, and S. B. Haase. Global control of cell-cycle transcription by coupled CDK and network oscillators. *Nature*, 453(7197):944–947, 2008.

[282] A. Oshlack, M. D. Robinson, and M. D. Young. From RNA-seq reads to differential expression results. *Genome Biology*, 11(12), 2010.

[283] R. D. M. Page. Comments on component compatibility in historical biogeography. *Cladistics*, 5:167–182, 1989.

[284] R. D. M. Page. TBMap: a taxonomic perspective on the phylogenetic database TreeBASE. *BMC Bioinformatics*, 8, 2007.

[285] R. D. M. Page and M. A. Charleston. Trees within trees: Phylogeny and historical associations. *Trends in Ecology and Evolution*, 13:356–359, 1998.

[286] R. D. M. Page and E. C. Holmes. *Molecular Evolution: A Phylogenetic Approach*. Blackwell Scientific, 1998.

[287] S. W. Park, Y. Park, and M. I. Kuroda. Regulation of histone H4 Lys16 acetylation by predicted alternative secondary structures in *roX* noncoding RNAs. *Molecular and Cellular Biology*, 28(16):4952–4962, 2008.

[288] S. W. Park, I. K. Yool, J. G. Sypula, J. Choi, H. Oh, and Y. Park. An evolutionarily conserved domain of *roX2* RNA is sufficient for induction of H4-Lys16 acetylation on the *Drosophila* X chromosome. *Genetics*, 177(3):1429–1437, 2007.

[289] Y. Park, R. L. Kelley, H. Oh, M. I. Kuroda, and V. H. Meller. Extent of chromatin spreading determined by *roX* RNA recruitment of MSL proteins. *Science*, (5598):1620–1623, 2002.

[290] Y. Park and M. I. Kuroda. Epigenetic aspects of X-chromosome dosage compensation. *Science*, (5532), 2001.

[291] N. Patel and J. T. L. Wang. Semi-supervised prediction of gene regulatory networks using machine learning algorithms. *Journal of Biosciences*, 40(4):731–740, 2015.

[292] B. Paten, M. Diekhans, N. Gassner, W. James Kent, P. Mantey, A. M. Novak, M. Smuga-Otto, J. M. Stuart, D. Haussler, B. J. Druker, S. Friend, J. Guinney, M. Guttman, B. Wold, A. A. Margolin, M. Massie, F. Nothaft, D. Patterson, L. Pachter, and L. Van't Veer. The NIH BD2K center for big data in translational genomics. *Journal of the American Medical Informatics Association*, 22(6):1143–1147, 2015.

[293] J. Pearl. *Causality: Models, Reasoning, and Inference*. Cambridge University Press, 2000.

[294] D. Pe'er and N. Hacohen. Principles and strategies for developing network models in cancer. *Cell*, 144(6):864–873, 2011.

[295] L. Peng, L. K. Ng, and S. See. YellowRiver: A flexible high performance cluster computing service for grid. In *Proceedings - Eighth International Conference on High-Performance Computing in Asia-Pacific Region*, volume 2005, pages 553–558, Nanyang, 2005.

[296] W. H. Piel, M. J. Sanderson, and M. J. Donoghue. The small-world dynamics of tree networks and data mining in phyloinformatics. *Bioinformatics*, 19(9):1162–1168, 2003.

[297] C. W. A. Pleij, K. Rietveld, and L. Bosch. A new principle of RNA folding based on pseudoknotting. *Nucleic Acids Research*, 13(5):1717–1731, 1985.

[298] K. Prüfer, S. Sawyer, A. Heinze, G. Renaud, C. De Filippo, M. Dannemann, Q. Fu, M. Kircher, M. Kuhlwilm, M. Lachmann, M. Meyer, M. Ongyerth, M. Siebauer, C. Theunert, J. Kelso, S. Pääbo, F. Racimo, F. Jay, M. Slatkin, N. Patterson, S. Sankararaman, H. Li, S. Mallick, A. Tandon, D. Reich, P. Moorjani, J. Pickrell, P. H. Sudmant, J. O. Kitzman, J. Shendure, E. E. Eichler, J. C. Mullikin, S. H. Vohr, R. E. Green, I. Hellmann, P. L. F. Johnson, H. Blanche, H. Cann, E. S. Lein, T. E. Bakken, L. V. Golovanova, V. B. Doronichev, M. V. Shunkov, A. P. Derevianko, and B. Viola. The complete genome sequence of a Neanderthal from the Altai Mountains. *Nature*, 505(7481):43–49, 2014.

[299] M. Ptashne and A. Gann. *Genes & Signals*. Cold Spring Harbor Laboratory Press, 2002.

[300] T. Puton, L. P. Kozlowski, K. M. Rother, and J. M. Bujnicki. CompaRNA: a server for continuous benchmarking of automated methods for RNA secondary structure prediction. *Nucleic Acids Research*, 41(7):4307–4323, 2013.

[301] A. M. Pyle and Z. Shakked. The ever-growing complexity of nucleic acids: from small DNA and RNA motifs to large molecular assemblies and machines (editorial overview). *Current Opinions in Structural Biology*, 21:293–295, 2011.

[302] E. Rahm and H. H. Do. Data cleaning: problems and current approaches. *Bulletin of Technical Committee on Data Engineering, Special Issue on Data Cleaning*, 23(4):3–13, 2000.

[303] R. R. Rahrig, N. B. Leontis, and C. L. Zirbel. R3D Align: global pairwise alignment of RNA 3D structures using local superpositions. *Bioinformatics*, 26(21):2689–2697, 2010.

[304] R. R. Rahrig, A. I. Petrov, N. B. Leontis, and C. L. Zirbel. R3D Align Web server for global nucleotide to nucleotide alignments of RNA 3D structures. *Nucleic Acids Research*, 41:W15–W21, 2013.

[305] V. Raman and J. M. Hellerstein. Potter's wheel: an interactive data cleaning system. In *Proceedings of 27th International Conference on Very Large Data Bases*, pages 381–390, 2001.

[306] J. Reichert and J. Suhnel. The IMB Jena image library of biological macromolecules: 2002 update. *Nucleic Acids Research*, 30:253–254, 2002.

[307] C. M. Reidys, F. W. D. Huang, J. E. Andersen, R. C. Penner, P. F. Stadler, and M. E. Nebel. Topology and prediction of RNA pseudoknots. *Bioinformatics*, 27(8):1076–1085, 2011.

[308] N. J. Reiter, C. W. Chan, and A. Mondrago. Emerging structural themes in large RNA molecules. *Current Opinions in Structural Biology*, 21:319–326, 2011.

[309] J. S. Reuter and D. H. Mathews. RNAstructure: software for RNA secondary structure prediction and analysis. *BMC Bioinformatics*, 11, 2010.

[310] J. A. Robles, S. J. Stephen, J. M. Taylor, S. E. Qureshi, S. R. Wilson, and C. J. Burden. Efficient experimental design and analysis strategies for the detection of differential expression using RNA-sequencing. *BMC Genomics*, 13(1), 2012.

[311] U. Roshan and D. R. Livesay. Probalign: multiple sequence alignment using partition function posterior probabilities. *Bioinformatics (Oxford, England)*, 22(22):2715–2721, 2006.

[312] S. Roy, J. Ernst, P. Kheradpour, C. A. Bristow, M. F. Lin, S. Washietl, F. Ay, P. E. Meyer, L. Di Stefano, R. Candeias, I. Jungreis, D. Marbach, R. Sealfon, S. Will, B. Berger, M. Kellis, P. V. Kharchenko, M. Y. Tolstorukov, Y. B. Schwartz, P. J. Park, N. Negre, L. Ma, C. D. Brown, R. Spokony, R. Grossman, K. P. White, M. L. Eaton, H. K. MacAlpine, S. K. Powell, D. M. MacAlpine, J. M. Landolin, J. W. Carlson, B. W. Booth, A. Minoda, J. E. Sandler, K. H. Wan, C. Yu, R. A. Hoskins, S. E. Celniker, G. H. Karpen, B. I. Arshinoff, N. Robine, Q. Dai, J. G. Henikoff, K. Okamura, E. C. Lai, N. L. Washington, S. Lewis, A. Carr, S. Russell, G. Micklem, A. A. Alekseyenko, A. A. Gorchakov, M. I. Kuroda, C. Artieri, J. Malone, D. Sturgill, B. Oliver, C. A. Davis, X. Feng, T. R. Gingeras, M. O. Duff, L. Yang, B. R. Graveley, T. Gu, N. C. Riddle, S. C. R. Elgin, P. Kapranov, R. Li, E. Feingold, P. Good, M. Guyer, R. Lowdon, M. Perry, L. D. Stein, A. Samsonova, N. Perrimon, M. van Baren, M. R. Brent, J. Andrews, L. Cherbas, T. C. Kaufman, P. Cherbas, J. Nordman, N. Sher, T. Orr-Weaver, E. Berezikov, S. Henikoff, J. W. Posakony, B. Ren, V. Pirrotta, A. N. Brooks, S. E. Brenner, A. Sakai, K. Ahmad, W. Kent, B. Booth, A. A. Samsonova, M. van Baren, T. R. Gingeras, R. A. Hoskins, C. L. G. Comstock, A. Dobin, J. Drenkow, S. Dudoit, J. Dumais, D. Fagegaltier, S. Ghosh, K. D. Hansen, S. Jha, L. Langton, W. Lin, D. Miller, A. E. Tenney, H. Wang, A. T. Willingham, C. Zaleski, D. Zhang, D. Acevedo, E. P. Bishop, S. E. Gadel, Y. L. Jung, C. D. Kennedy, O. K. Lee, D. Linder-Basso, S. E. Marchetti, G. Shanower, N. Nègre, R. L. Grossman, R. Auburn, H. J. Bellen, J. Chen, M. H. Domanus, D. Hanley, E. Heinz, Z. Li, F. Meyer, S. W. Miller, C. A. Morrison, D. A. Scheftner, L. Senderowicz, P. K. Shah, P. Suchy, F. Tian, K. J. T. Venken, R. White, J. Wilkening, J. Zieba, J. T. Nordman, T. L. Orr-Weaver, L. C. DeNapoli, Q. Ding, T. Eng, H. Kashevsky, S. Li, J. A. Prinz, Q. Dai, G. J. Hannon, M. Hirst, M. Marra, M. Rooks, Y. Zhao, T. D. Bryson, M. D. Perry, W. J. Kent, S. E. Lewis, G. Barber, A. Chateigner, H. Clawson, S. Contrino, F. Guillier, A. S. Hinrichs, E. T. Kephart, P. Lloyd, R. Lyne, S. McKay, R. A. Moore, C. Mungall, K. M. Rutherford, P. Ruzanov, R. Smith, E. O. Stinson, Z. Zha, C. G. Artieri, J. H. Malone, L. Jiang, N. Mattiuzzo, E. A. Feingold, P. J. Good, M. S. Guyer, and R. F. Lowdon. Identification of functional elements and regulatory circuits

by *Drosophila* modENCODE. *Science*, 330(6012):1787–1797, 2010.

[313] Y. Saeys, I. Inza, and P. Larrañaga. A review of feature selection techniques in bioinformatics. *Bioinformatics*, 23(19):2507–2517, 2007.

[314] R. Saito, M. E. Smoot, K. Ono, J. Ruscheinski, P. L. Wang, T. Ideker, S. Lotia, A. R. Pico, and G. D. Bader. A travel guide to Cytoscape plugins. *Nature Methods*, 9(11):1069–1076, 2012.

[315] N. Saitou and M. Nei. The neighbor-joining method: a new method for reconstructing phylogenetic trees. *Molecular Biology and Evolution*, 4:406–425, 1987.

[316] H. Salgado, M. Peralta-Gil, S. Gama-Castro, A. Santos-Zavaleta, L. Muñiz Rascado, J. S. García-Sotelo, V. Weiss, H. Solano-Lira, I. Martínez-Flores, A. Medina-Rivera, G. Salgado-Osorio, S. Alquicira-Hernández, K. Alquicira-Hernández, A. López-Fuentes, L. Porrón-Sotelo, A. M. Huerta, C. Bonavides-Martínez, Y. I. Balderas-Martínez, L. Pannier, V. Del Moral-Chávez, A. Hernández-Alvarez, J. Collado-Vides, M. Olvera, A. Labastida, E. Morett, V. Jiménez-Jacinto, and L. Vega-Alvarado. RegulonDB v8.0: Omics data sets, evolutionary conservation, regulatory phrases, cross-validated gold standards and more. *Nucleic Acids Research*, 41(D1):D203–D213, 2013.

[317] J. Sanders and E. Kandrot. *CUDA by Example: An Introduction to General-Purpose GPU Programming*. Addison-Wesley, 2011.

[318] M. J. Sanderson, M. J. Donoghue, W. H. Piel, and T. Eriksson. TreeBASE: a prototype database of phylogenetic analyses and an interactive tool for browsing the phylogeny of life. *American Journal of Botany*, 81(6), 1994.

[319] M. J. Sanderson, A. Purvis, and C. Henze. Phylogenetic supertrees: assembling the trees of life. *Trends in Ecology & Evolution*, 13.3:105–109, 1998.

[320] K. Sato, Y. Kato, T. Akutsu, K. Asai, and Y. Sakakibara. DAFS: simultaneous aligning and folding of RNA sequences via dual decomposition. *Bioinformatics*, 28(24):3218–3224, 2012.

[321] S. M. Savaresi, D. L. Boley, S. Bittanti, and G. Gazzaniga. Cluster selection in divisive clustering algorithms. In *SIAM International Conference on Data Mining*, pages 299–314, 2002.

[322] T Schaffter, D Marbach, and D Floreano. GeneNetWeaver: in silico benchmark generation and performance profiling of network inference methods. *Bioinformatics*, 27:2263–70, 2011.

[323] M. C. Schatz. CloudBurst: highly sensitive read mapping with MapReduce. *Bioinformatics*, 25(11):1363–1369, 2009.

[324] F. Schluenzen, A. Tocilj, J. Harms, M. Gluehmann, D. Janell, A. Yonath, R. Zarivach, A. Bashan, H. Bartels, I. Agmon, and F. Franceschi. Structure of functionally activated small ribosomal subunit at 3.3 Å resolution. *Cell*, 102(5):615–623, 2000.

[325] H. Schmidt and M. Jirstrand. Systems biology toolbox for MAT-LAB: A computational platform for research in systems biology. *Bioinformatics*, 22(4):514–515, 2006.

[326] G. D. Schuler, J. A. Epstein, H. Ohkawa, and J. A. Kans. Entrez: molecular biology database and retrieval system. *Methods in Enzymology*, 266:141–162, 1996.

[327] Y. Senbabaoğlu, S. O. Sümer, F. Sánchez-Vega, D. Bemis, G. Ciriello, N. Schultz, and C. Sander. A multi-method approach for proteomic network inference in 11 human cancers. *PLoS Computational Biology*, 12(2), 2016.

[328] O. Shalem, O. Dahan, M. Levo, M. R. Martinez, I. Furman, E. Segal, and Y. Pilpel. Transient transcriptional responses to stress are generated by opposing effects of mRNA production and degradation. *Molecular Systems Biology*, 4, 2008.

[329] H. Shan. *PhD Dissertation: An Approximate Search Engine for Structure*. PhD thesis, New Jersey Institute of Technology, Newark, 2004.

[330] C. E. Shannon. A mathematical theory of communication. *Bell System Technical Journal*, 27(3):379–423, July 1948.

[331] B. A. Shapiro and K. Zhang. Comparing multiple RNA secondary structures using tree comparisons. *Computer Applications in the Biosciences*, 6(4):309–318, 1990.

[332] M. J. Sharkey and J. W. Leathers. Majority does not rule: the trouble with majority-rule consensus trees. *Cladistics*, 17(3):282–284, 2001.

[333] D. Shasha, J. T. L. Wang, H. Shan, and K. Zhang. ATreeGrep: approximate searching in unordered trees. In *Proceedings of 14th International Conference on Scientific and Statistical Database Management*, pages 89–98, 2002.

[334] C. Shearer. The CRISP-DM model: the new blueprint for data mining. *Journal of Data Warehousing*, 5(4), 2000.

[335] L. X. Shen and I. Tinoco Jr. The structure of an RNA pseudoknot that causes efficient frameshifting in mouse mammary tumor virus. *Journal of Molecular Biology*, 247(5):963–978, 1995.

[336] A. C. G. Silveira, F. L. Thompson, A. T. R. Vasconcelos, K. L. Robertson, B. Lin, Z. Wang, and G. J. Vora. Identification of non-coding RNAs in environmental *Vibrios*. *Microbiology*, 156(8):2452–2458, 2010.

[337] S. Smit, J. Heringa, K. Rother, and R. Knight. From knotted to nested RNA structures: a variety of computational methods for pseudoknot removal. *RNA*, 14(3):410–416, 2008.

[338] B. Smith, M. Ashburner, C. Rosse, J. Bard, W. Bug, W. Ceusters, L. J. Goldberg, K. Eilbeck, A. Ireland, P. Rocca-Serra, S. A. Sansone, C. J. Mungall, S. Lewis, N. Leontis, A. Ruttenberg, R. H. Scheuermann, N. Shah, and P. L. Whetzel. The OBO Foundry: coordinated evolution of ontologies to support biomedical data integration. *Nature Biotechnology*, 25(11):1251–1255, 2007.

[339] C. Soneson and M. Delorenzi. A comparison of methods for differential expression analysis of RNA-seq data. *BMC Bioinformatics*, 14(1), 2013.

[340] Y. Song. *PhD Dissertation: Data Mining in Computational Proteomics and Genomics*. PhD thesis, New Jersey Institute of Technology, Newark, 2015.

[341] Y. Song, L. Hua, B. A. Shapiro, and J. T. L. Wang. Effective alignment of RNA pseudoknot structures using partition function posterior log-odds scores. *BMC Bioinformatics*, 16(1), 2015.

[342] Y. Song, C. Liu, R. Malmberg, F. Pan, and L. Cai. Tree decomposition-based fast search of RNA structures including pseudoknots in genomes. In *Proceedings of IEEE Computational Systems Bioinformatics Conference*, volume 1, pages 223–234, 2005.

[343] E. L. L. Sonnhammer, S. R. Eddy, and R. Durbin. Pfam: a comprehensive database of protein domain families based on seed alignments. *Proteins: Structure, Function and Genetics*, 28:405–442, 1998.

[344] D. A. Sorescu, M. Möhl, M. Mann, R. Backofen, and S. Will. CARNA alignment of RNA structure ensembles. *Nucleic Acids Research*, 40(W1):W49–W53, 2012.

[345] D. W. Staple and S. E. Butcher. Pseudoknots: RNA structures with diverse functions. *PLoS Biology*, 3(6), 2005.

[346] A. Stark, M. F. Lin, M. Kellis, P. Kheradpour, M. D. Rasmussen, A. N. Deoras, J. S. Pedersen, A. S. Hinrichs, B. Paten, W. J. Kent, D. Haussler, L. Parts, J. W. Carlson, S. E. Celniker, C. Yu, S. Park, K. H. Wan, M. A. Crosby, W. M. Gelbart, B. B. Matthews, A. J. Schroeder, L. S. Gramates, S. E. St Pierre, M. Roark, K. L. Wiley Jr., R. J. Kulathinal, P. Zhang, K. V. Myrick, J. V. Antone, S. Roy, J. G. Ruby, D. P. Bartel, J. Brennecke, E. Hodges, G. J. Hannon, A. Caspi, S. W. Park, Y. Park, M. V. Han, M. W. Hahn, M. L. Maeder, B. J. Polansky, B. E. Robson, D. A. Eastman, S. Aerts, B. Hassan, J. Van Helden, D. G. Gilbert, T. C. Kaufman, M. Rice, M. Weir, C. N. Dewey, L. Pachter, E. C. Lai, M. B. Eisen, A. G. Clark, and D. Smith. Discovery of functional elements in 12 *Drosophila* genomes using evolutionary signatures. *Nature*, 450(7167):219–232, 2007.

[347] R. Stevens, C. Goble, N. Paton, S. Bechhofer, G. Ng, P. Baker, and A. Brass. Complex query formulation over diverse information sources in TAMBIS. In *Bioinformatics: Managing Scientific Data*, pages 189–224. Morgan Kaufmann, 2003.

[348] C. Stockham, L. S. Wang, and T. Warnow. Statistically based postprocessing of phylogenetic analysis by clustering. *Bioinformatics*, 18 Suppl 1:S285–S293, 2002.

[349] C. J. Stoeckert Jr., F. Salas, B. Brunk, and G. C. Overton. EpoDB: a prototype database for the analysis of genes expressed during vertebrate erythropoiesis. *Nucleic Acids Research*, 27(1):200–203, 1999.

[350] G. Stoesser, W. Baker, A. van den Broek, M. Garcia-Pastor, C. Kanz, T. Kulikova, R. Leinonen, Q. Lin, V. Lombard, R. Lopez, R. Mancuso, F. Nardone, P. Stoehr, M. A. Tali, K. Tzouvara, and R. Vaughan. The EMBL nucleotide sequence database: major new

developments in nucleic acids research. *Nucleic Acids Research*, 311:17–22, 2003.

[351] G. Storz. An expanding universe of noncoding RNAs. *Science*, (5571), 2002.

[352] C. Stuckenholz, M. I. Kuroda, and V. H. Meller. Functional redundancy within *roX1*, a noncoding RNA involved in dosage compensation in *Drosophila melanogaster*. *Genetics*, 164(3):1003–1014, 2003.

[353] V. A. Stupina, A. Meskauskas, J. C. McCormack, J. D. Dinman, A. E. Simon, Y. G. Yingling, and B. A. Shapiro. The 3′ proximal translational enhancer of Turnip crinkle virus binds to 60S ribosomal subunits. *RNA*, 14(11):2379–2393, 2008.

[354] A. R. Subramanian, M. Kaufmann, and B. Morgenstern. DIALIGN-TX: Greedy and progressive approaches for segment-based multiple sequence alignment. *Algorithms for Molecular Biology*, 3(1), 2008.

[355] Y. Sun, J. Buhler, and C. Yuan. Designing filters for fast-known ncRNA identification. *IEEE/ACM Transactions on Computational Biology and Bioinformatics*, 9(3):774–787, 2012.

[356] T. Suzuki, M. Sugiyama, J. Sese, and T. Kanamori. Approximating mutual information by maximum likelihood density ratio estimation. In *Journal of Machine Learning Research; Workshop and Conference Proceedings*, volume 4, pages 5–20, 2008.

[357] Y. Tabei, K. Asai, H. Kiryu, and T. Kin. A fast structural multiple alignment method for long RNA sequences. *BMC Bioinformatics*, 9, 2008.

[358] P. Tan, M. Steinbach, and V. Kumar. *Introduction to Data Mining*. Addison Wesley, 2006.

[359] S. Tarazona, F. García-Alcalde, J. Dopazo, A. Conesa, and A. Ferrer. Differential expression in RNA-seq: a matter of depth. *Genome Research*, 21(12):2213–2223, 2011.

[360] Y. Tateno, K. Fukami-Kobayashi, S. Miyazaki, H. Sugawara, and T. Gojobori. DNA Data Bank of Japan at work on genome sequence data. *Nucleic Acids Research*, 26(1):16–20, 1998.

[361] M. Taufer, A. Licon, R. Araiza, D. Mireles, M. Y. Leung, F. H. D. van Batenburg, and A. P. Gultyaev. PseudoBase++: an extension

of pseudobase for easy searching, formatting and visualization of pseudoknots. *Nucleic Acids Research*, 37(1):D127–D135, 2009.

[362] Consortium of The *C. elegans* Sequencing. Genome sequence of the nematode *C. elegans*: a platform for investigating biology. *Science*, (5396), 1998.

[363] C. A. Theimer, C. A. Blois, and J. Feigon. Structure of the human telomerase RNA pseudoknot reveals conserved tertiary interactions essential for function. *Molecular Cell*, 17(5):671–682, 2005.

[364] J. D. Thompson, F. Plewniak, and O. Poch. A comprehensive comparison of multiple sequence alignment programs. *Nucleic Acids Research*, 27(13):2682–2690, 1999.

[365] J. L. Thorley and R. D. M. Page. RadCon: phylogenetic tree comparison and consensus. *Bioinformatics*, 16:486–487, 2000.

[366] N. Toor, K. S. Keating, S. D. Taylor, and A. M. Pyle. Crystal structure of a self-spliced group II intron. *Science*, 320:77–82, 2008.

[367] E. Torarinsson, J. H. Havgaard, and J. Gorodkin. Multiple structural alignment and clustering of RNA sequences. *Bioinformatics*, 23(8):926–932, 2007.

[368] C. Trapnell, S. L. Salzberg, and L. Pachter. TopHat: discovering splice junctions with RNA-seq. *Bioinformatics*, 25(9):1105–1111, 2009.

[369] M. F. Traxler, D.-E. Chang, and T. Conway. Guanosine 3′,5′-bispyrophosphate coordinates global gene expression during glucose-lactose diauxie in *Escherichia coli*. *Proceedings of National Academy of Sciences of the United States of America*, (7), 2006.

[370] F. H. D. van Batenburg, A. P. Gultyaev, C. W. A. Pleij, J. Ng, and J. Oliehoek. PseudoBase: a database with RNA pseudoknots. *Nucleic Acids Research*, 28(1):201–204, 2000.

[371] V. N. Vapnik and V. Vapnik. *Statistical Learning Theory*, volume 1. John Wiley & Sons, 1998.

[372] P. Vassiliadis, Z. Vagena, S. Skiadopoulos, N. Karayannidis, and T. K. Sellis. ARKTOS: towards the modeling, design, control and execution of ETL processes. *Information Systems*, 26(8):537–561, 2001.

[373] V. E. Velculescu, L. Zhang, B. Vogelstein, and K. W. Kinzler. Serial analysis of gene expression. *Science*, 270:484–487, 1995.

[374] M. Vignes, J. Vandel, D. Allouche, N. Ramadan-Alban, C. Cierco-Ayrolles, T. Schiex, B. Mangin, and S. de Givry. Gene regulatory network reconstruction using Bayesian networks, the Dantzig selector, the Lasso and their meta-analysis. *PLoS One*, 6(12), 2011.

[375] A. E. Walter, D. H. Turner, J. Kim, M. H. Lyttle, P. Müller, D. H. Mathews, and M. Zuker. Coaxial stacking of helixes enhances binding of oligoribonucleotides and improves predictions of RNA folding. *Proceedings of National Academy of Sciences of the United States of America*, 91(20):9218–9222, 1994.

[376] A. X. Wang, W. L. Ruzzo, and M. Tompa. How accurately is ncRNA aligned within whole-genome multiple alignments?. *BMC Bioinformatics*, 8, 2007.

[377] Chih-Wei Wang, Kun-Tze Chen, and Chin Lung Lu. iPARTS: an improved tool of pairwise alignment of RNA tertiary structures. *Nucleic Acids Research*, 38:W340–W347, 2010.

[378] J. Wang, W. Wang, R. Li, Y. Li, G. Tian, L. Goodman, W. Fan, J. Zhang, J. Li, J. Zhang, Y. Guo, B. Feng, H. Li, Y. Lu, X. Fang, H. Liang, Z. Du, D. Li, Y. Zhao, Y. Hu, Z. Yang, H. Zheng, X. Yi, J. Zhao, J. Duan, Y. Zhou, J. Qin, L. Ma, G. Li, Z. Yang, G. Zhang, B. Yang, C. Yu, F. Liang, W. Li, S. Li, D. Li, P. Ni, J. Ruan, Q. Li, H. Zhu, D. Liu, Z. Lu, N. Li, G. Guo, J. Zhang, J. Ye, L. Fang, Q. Hao, Q. Chen, Y. Liang, and Y. Su. The diploid genome sequence of an Asian individual. *Nature*, 456(7218):60–65, 2008.

[379] J. T. L. Wang, H. Shan, D. Shasha, and W. H. Piel. TreeRank: a similarity measure for nearest neighbor searching in phylogenetic databases. In *Proceedings of 15th International Conference on Scientific and Statistical Database Management*, pages 171–180, 2003.

[380] J. T. L. Wang, H. Shan, D. Shasha, and W. H. Piel. Fast structural search in phylogenetic databases. *Evolutionary Bioinformatics Online*, 1:37–46, 2007.

[381] J. T. L. Wang, M. J. Zaki, H. T. T. Toivonen, and D. Shasha (eds.). *Data Mining in Bioinformatics*. Springer, 2005.

[382] J. T. L. Wang, K. Zhang, G. Chang, and D. Shasha. Finding approximate patterns in undirected acyclic graphs. *Pattern Recognition*, 35(2):473–483, 2002.

[383] J. T. L. Wang, K. Zhang, K. Jeong, and D. Shasha. A system for approximate tree matching. *IEEE Transactions on Knowledge and Data Engineering*, 6(4):559–571, 1994.

[384] L. Wang and D. Gusfield. Constructing additive trees when the error is small. *Journal of Computational Biology*, 5(1):127–134, 1998.

[385] L. Wang, T. Jiang, and D. Gusfield. A more efficient approximation scheme for tree alignment. *SIAM Journal on Computing*, 30(1):283–299, 2000.

[386] Y. Wang, H. Zhou, N. Ghaffari, C. D. Johnson, U. M. Braga-Neto, H. Wang, and R. Chen. Evaluation of the coverage and depth of transcriptome by RNA-seq in chickens. *BMC Bioinformatics*, 12(10), 2011.

[387] Z Wang, M Gerstein, and M Snyder. RNA-seq: a revolutionary tool for transcriptomics. *Nature Reviews Genetics*, 10:57–63, 2009.

[388] O. Weichenrieder, K. Wild, S. Cusack, and K. Strub. Structure and assembly of the Alu domain of the mammalian signal recognition particle. *Nature*, 408(6809):167–173, 2000.

[389] Z. Weinberg and W. L. Ruzzo. Sequence-based heuristics for faster annotation of non-coding RNA families. *Bioinformatics*, 22(1):35–39, 2006.

[390] J. Westbrook, Z. Feng, L. Chen, H. Yang, and H. M. Berman. The Protein Data Bank and structural genomics. *Nucleic Acids Research*, 31:489–491, 2003.

[391] J. Westbrook, Z. Feng, S. Jain, T. N. Bhat, N. Thanki, V. Ravichandran, G. L. Gilliland, W. Bluhm, H. Weissig, D. S. Greer, P. E. Bourne, and H. M. Berman. The Protein Data Bank: unifying the archive. *Nucleic Acids Research*, 30:245–248, 2002.

[392] D. L. Wheeler, D. M. Church, A. E. Lash, D. D. Leipe, T. L. Madden, J. U. Pontius, G. D. Schuler, L. M. Schriml, T. A. Tatusova, L. Wagner, and B. A. Rapp. Database resources of the National Center for Biotechnology Information. *Nucleic Acids Research*, 30:13–16, 2002.

[393] F. Wilcoxon. Probability tables for individual comparisons by ranking methods. *Biometrics*, (3), 1947.

[394] S. Will, R. Backofen, K. Reiche, P. F. Stadler, and I. L. Ho-facker. Inferring noncoding RNA families and classes by means of genome-scale structure-based clustering. *PLoS Computational Biology*, 3(4):680–691, 2007.

[395] S. Will, T. Joshi, R. Backofen, I. L. Hofacker, and P. F. Stadler. LocARNA-P: accurate boundary prediction and improved detection of structural RNAs. *RNA*, 18(5):900–914, 2012.

[396] W. T. Williams and H. T. Clifford. On the comparison of two classifications on the same set of elements. *Taxon*, 20:519–522, 1971.

[397] A. Wilm, I. Mainz, and G. Steger. An enhanced RNA alignment benchmark for sequence alignment programs. *Algorithms for Molecular Biology*, 1, 2006.

[398] B. T. Wimberly, D. E. Brodersen, W. M. Clemons Jr., R. J. Morgan-Warren, A. P. Carter, V. Ramakrishnan, C. Vonrheln, and T. Hartsch. Structure of the 30S ribosomal subunit. *Nature*, 407(6802):327–339, 2000.

[399] T. K. Wong, T. W. Lam, W. K. Sung, and S. M. Yiu. Adjacent nucleotide dependence in ncRNA and order-1 SCFG for ncRNA identification. *PLoS One*, 5(9), 2010.

[400] T. K. F. Wong, K. L. Wan, B. W. Y. Cheung, T. W. Lam, S. M. Yiu, B. Y. Hsu, and W. K. Hon. RNASAlign: RNA structural alignment system. *Bioinformatics*, 27(15):2151–2152, 2011.

[401] C. H. Wu, R. Apweiler, A. Bairoch, D. A. Natale, W. C. Barker, B. Boeckmann, S. Ferro, E. Gasteiger, H. Huang, R. Lopez, M. Magrane, M. J. Martin, R. Mazumder, C. O'Donovan, N. Redaschi, and B. Suzek. The Universal Protein Resource (UniProt): an expanding universe of protein information. *Nucleic Acids Research*, 34:D187–D191, 2006.

[402] C. H. Wu, H. Huang, A. Nikolskaya, Z. Z. Hu, and W. C Barker. The iProClass integrated database for protein functional analysis. *Computational Biology and Chemistry*, 28:87–96, 2004.

[403] C. H. Wu, A. Nikolskaya, H. Huang, L. L. Yeh, D. A. Natale, C. R. Vinayaka, Z. Hu, R. Mazumder, S. Kumar, P. Kourtesis, R. S. Ledley, B. E. Suzek, L. Arminski, Y. Chen, J. Zhang, J. L. Cardenas, S. Chung, J. Castro-Alvear, G. Dinkov, and W. C. Barker. PIRSF: family classification system at the Protein Information Resource. *Nucleic Acids Research*, 32:D112–D114, 2004.

[404] C. H. Wu, L. S. Yeh, H. Huang, L. Arminski, J. Castro-Alvear, Y. Chen, Z. Z. Hu, R. S. Ledley, P. Kourtesis, B. E. Suzek, C. R. Vinayaka, J. Zhang, and W. C. Barker. The protein information resource. *Nucleic Acids Research*, 31:345–347, 2003.

[405] Y. Xin, C. Laing, N. B. Leontis, and T. Schlick. Annotation of tertiary interactions in RNA structures reveals variations and correlations. *RNA*, 14:2465–2477, 2008.

[406] H. Xu, C. Baroukh, R. Dannenfelser, E. Y. Chen, C. M. Tan, Y. Kou, Y. E. Kim, I. R. Lemischka, and A. Ma'ayan. ESCAPE: database for integrating high-content published data collected from human and mouse embryonic stem cells. *Database: The Journal of Biological Databases & Curation*, 2013:1–12, 2013.

[407] X. Xu, Y. Ji, and G. D. Stormo. RNA Sampler: a new sampling-based algorithm for common RNA secondary structure prediction and structural alignment. *Bioinformatics*, 23(15):1883–1891, 2007.

[408] H. Yang, F. Jossinet, N. Leontis, L. Chen, J. Westbrook, H. Berman, and E. Westhof. Tools for the automatic identification and classification of RNA base pairs. *Nucleic Acids Research*, 31(13):3450–3460, 2003.

[409] Z. Yao, Z. Weinberg, and W.L. Ruzzo. CMfinder: a covariance model based RNA motif finding algorithm. *Bioinformatics*, 22(4):445–452, 2006.

[410] A. H. Yona, Y. S. Manor, R. H. Herbst, G. H. Romano, A. Mitchell, M. Kupiec, Y. Pilpel, and O. Dahan. Chromosomal duplication is a transient evolutionary solution to stress. *Proceedings of National Academy of Sciences of the United States of America*, (51), 2012.

[411] I. Yoo, P. Alafaireet, M. Marinov, K. Pena-Hernandez, R. Gopidi, J. F. Chang, and L. Hua. Data mining in healthcare and biomedicine: a survey of the literature. *Journal of Medical Systems*, 36:2431–2448, 2012.

[412] B. J. Yoon. Efficient alignment of RNAs with pseudoknots using sequence alignment constraints. *EURASIP Journal on Bioinformatics and Systems Biology*, 2009.

[413] W. C. Young, A. E. Raftery, and K. Y. Yeung. Fast Bayesian inference for gene regulatory networks using ScanBMA. *BMC Systems Biology*, 8(1), 2014.

[414] M. Zaharia, M. Chowdhury, T. Das, A. Dave, J. Ma, M. McCauley, M. J. Franklin, S. Shenker, and I. Stoica. Resilient distributed datasets: a fault-tolerant abstraction for in-memory cluster computing. In *Proceedings of 9th USENIX Conference on Networked Systems Design and Implementation*, Berkeley, 2012.

[415] K. A. Zaretskii. Constructing trees from the set of distances between pendant vertices. *Uspehi Matematicekih Nauk*, 20:90–92, 1965.

[416] S. Zhao. Assessment of the impact of using a reference transcriptome in mapping short RNA-seq reads. *PLoS One*, 9(7), 2014.

[417] S. Zhao, K. Prenger, and L. Smith. Stormbow: A cloud-based tool for read mapping and expression quantification in large-scale RNA-seq studies. *ISRN Bioinformatics*, 2013.

[418] S. Zhao, K. Prenger, L. Smith, T. Messina, H. Fan, E. Jaeger, and S. Stephens. Rainbow: a tool for large-scale whole-genome sequencing data analysis using cloud computing. *BMC Genomics*, 14, 2013.

[419] C. Zhong and S. Zhang. Efficient alignment of RNA secondary structures using sparse dynamic programming. *BMC Bioinformatics*, 14, 2013.

[420] B. Zhu, G. Leroy, H. Chen, and Y. Chen. MedTextus: an intelligent web-based medical meta-search system. In *Proceedings of ACM International Conference on Digital Libraries*, Tuscon, 2002.

[421] C. M. Zmasek and S. R. Eddy. A simple algorithm to infer gene duplication and speciation events on a gene tree. *Bioinformatics*, 17:821–828, 2001.

[422] P. Zoppoli, S. Morganella, and M. Ceccarelli. TimeDelay-ARACNE: Reverse engineering of gene networks from time-course data by an information theoretic approach. *BMC Bioinformatics*, 11, 2010.

Index

Printed and bound by CPI Group (UK) Ltd, Croydon, CR0 4YY

23/10/2024

01777671-0012